65 Topics in Current Chemistry
Fortschritte der chemischen Forschung

Theoretical Inorganic Chemistry II

Springer-Verlag
Berlin Heidelberg GmbH 1976

This series presents critical reviews of the present position and future trends in modern chemical research. It is addressed to all research and industrial chemists who wish to keep abreast of advances in their subject.

As a rule, contributions are specially commissioned. The editors and publishers will, however, always be pleased to receive suggestions and supplementary information. Papers are accepted for "Topics in Current Chemistry" in either German or English.

ISBN 978-3-662-15980-4   ISBN 978-3-540-38188-4 (eBook)
DOI 10.1007/978-3-540-38188-4

Library of Congress Cataloging in Publication Data (Revised) Main entry under title: Theoretical inorganic chemistry. Includes bibliographies and index. 1. Chemistry, Physical and theoretical – Addresses, essays, lectures. I. Series. QD1.F58 vol. 56 540'.8s[541'.2] 75-5565

Originally published by Springer-Verlag Berlin Heidelberg New York in 1976.
Softcover reprint of the hardcover 1st edition 1976

Contents

Springer-Verlag	Postfach 105280 · D-6900 Heidelberg 1 Telephone (06221) 487-1 · Telex 04-61723 Heidelberger Platz 3 · D-1000 Berlin 33 Telephone (030) 822001 · Telex 01-83319
Springer-Verlag New York Inc.	New York, NY 10010 · 175, Fifth Avenue Telephone 673-2660

Diastereoisomerism and Diastereoselectivity in Metal Complexes

Prof. Dr. Klaus Bernauer

Université de Neuchâtel, Institut de Chimie, Neuchâtel, Suisse

Contents

1. Introduction

Systematic studies in the field of stereoselectivity of coordination compounds began rather late but have rapidly found a wide interest and several reviews have already appeared[1–7]. It is noteworthy that the problem of stereoselectivity in coordination compounds is very often related to the well-known stereochemical specificity of biological systems. Beyond the multiple functions that metal ions may assume in a metal enzyme, for example enhancement of reactivity through polarization or as a relay in electron transfers, they may also carry a stereochemical function. In a coordination sphere of definite geometry they can bring together the reacting molecules and increase not only the probability of the reaction but also orient them in a definite direction.

Unfortunately, the structure and the exact function of the metal ion is only known for a small number of metallo-enzymes. We may mention here as an example the carboxypeptidase – a zinc-enzyme which stereospecifically splits the terminal aminoacid from a peptide only in the case of the L-form.

The complexity of natural systems may be the reason why attempts were often made to reach a better understanding of these systems by the investigation of models. In doing this we should not overlook the fact that such models reflect only partial aspects in a very imperfect manner. With model systems, it has not been possible until now to obtain the extremely high catalytic effects and the high specificity of the corresponding systems. This indicates that the proposed models cannot simulate all the factors which determine the reactivity of natural systems. The molecular size of the latter may produce a net change in dielectric properties around the active site[8], which may be responsible for the stabilization of certain uncommon oxydation states of metal ions[9]. On the other hand, the metal ion is often found in a considerably distorted coordination sphere[10]. Therefore the results obtained by the investigation of a model may only answer those questions which are already included in the model.

In this context, the stereoselectivity seems to be an important element in the choice of a model. Its investigation will furnish broader information when it includes this element.

Another reason for model studies may be the desire to obtain catalytic and stereochemical effect in technical processes which are similar to those observed in natural systems.

One can define diastereoselectivity as the formation of diastereoisomers in a non-statistical ratio in any chemical transformation (formation of transition states included). Such a definition concerns equilibrium as well as nonreversible reactions. An asymmetric synthesis in a restricted sense can be considered as a reaction leading to a product containing at least one new stable dissymmetric center with a definite chirality. Such a reaction may take place in the coordination sphere of a metal ion. First of all the following question has to be answered: which are the structural properties in the architecture of the coordination sphere that lead to the following phenomenona:

a) high optical yield of a reaction going on in the coordination sphere and
b) a predictable chirality of the product?

In such an approach, kinetic and thermodynamic aspects and the position of the asymmetric element in the complex are of minor importance. So far the number of examples which show the properties mentioned above is very small. The optical yield in most examples of asymmetric synthesis in the coordination sphere of a metal is more or less an empirical result.

For an asymmetric reaction, two steps may be distinguished; they are schematically represented in Fig. 1

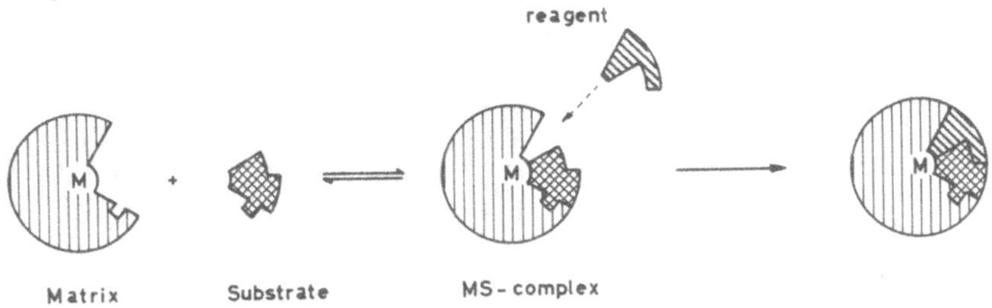

Fig. 1

In the first step, stereochemical information contained in a matrix-complex is transferred to the substrate by the formation of a matrix-substrate complex in such a way that the substrate in the second step will be able to differentiate the direction of attack of a reagent. In the reaction chain the MS-complex may appear as an intermediate or a transition state. The transition from the MS-complex to the product may proceed in a reversible or in a nonreversible reaction. Therefore the product ratio of an asymmetric synthesis is not necessarily controlled by kinetic parameters, although this may often be the case.

In discussing the appearence of diastereoselectivity in natural systems, the rule of three-point contact is frequently used[11]. This rule implies that a prochiral substrate becomes a chiral one, when it is fixed on three nonidentical points on the enzyme system. On the other hand, the well-known lock-key model is mainly based on a specific spacial arrangement and does not take into account specific types of bonding to the matrix[12]. For a discussion of diastereoselectivity in reactions with metal complexes, it seems more appropriate to use such a lock-key model.

Two distinct tendencies can be recognized in the literature dealing with diastereoselective reactions on metal complexes. On one side, stability differences are determined on the base of equilibrium data or product ratios. Inert cobalt(III)-complexes play a large part in this investigation. The procedures and the methods used in these investigations correspond to those already known in coordination chemistry. The aim of these studies is a better understanding of stereochemical interactions in the coordination sphere of metal complexes. The nature of ligand molecules is not modified in the systems studied. These reactions correspond to the first step of the scheme shown in Fig. 1.

On the other hand, we have the studies of reactions corresponding to the second step during which the chemical nature of a ligand molecule is modified. In such reactions a chiral complex is used as asymmetric agent. The problems arising in these investigations are mechanistic ones and are much more difficult to solve than those of equilibrium systems. The methods used are essentially those of organic analysis and synthesis.

It is quite clear that knowledge obtained from the study of equilibrium states may be applied to mechanistic problems only in a very restricted sense. Nevertheless it is surprising that the choice of chiral auxiliary reagents for synthesis seems seldom to be influenced by considerations of structural coordination chemistry. It must be mentioned that such reactions are often carried out in a medium not familiar to the coordination chemist.

The aim of following presentation is to discuss – in the case of known examples – the factors which may lead to a structure model which should be able to transmit chiral information in a reaction chain.

2. The Search for a Stereochemical Model

2.1. The Architecture of Coordination Compounds

The stereochemistry of a coordination compound is determined by

- the coordination number of the central metal atom
- the geometric arrangement of the ligand atoms
- the nature of ligand molecules

These factors include parameters such as bonding angles, bond length, etc. Isomers are compounds with identical composition but with a different stereochemical arrangement. We distinguish essentially between geometrical and optical isomers. The latter may be considered as a special case of the former. They are characterized by the presence of elements of definite chirality. In chelate compounds, four types of chiral elements may be distinguished:

- absolute configuration of a coordination entity with chiral arrangement of the chelate rings,
- conformation of a single chelate ring,
- asymmetric ligand atoms,
- optically active ligand molecules.

Diastereoisomerism is observed when two or more chiral elements appear in the same system. Isolation and characterization of enantiomers or diastereoisomers depend on the structural stability of the chiral element. Therefore the conformation of a chelate ring must generally be considered as labile. The structural stability of the configuration around the metal center and the stability of an asymmetric coordination atom depends on the nature of the central metal atom. In most cases an optically active ligand molecule can be considered as configurationally stable. The presence of a stable chiral element can induce definite chirality in another element which, taken separately, would be labile.

The most striking feature in diastereoisomerism of metal complexes is the very rapidly growing number of isomers, by introducing different elements of chirality in the basic framework of a coordination compound. The complex formation with 1,8-diamino-4-methyl-3,7-diazaoctane (5-Metrien) for example – a ligand with a single asymmetric center – leads not only to four geometric isomers. Three of these show a structure containing all four types of chiral elements mentioned, so that the system offers 28 possibilities of isomers.

One should not only classify different types of diastereoisomers according to the combination of chiral elements, but also different types of diastereoselectivity. Usually such a distinction is not made, which might often lead to some confusion. The formation of $[M(asp)_2]^-$ ($M = Ni^{2+}, Cu^{2+}$) shows no or only little selectivity considering a difference in stability between the optically active forms $[M(R\text{-}asp)_2]^-$ or $[M(S\text{-}asp)_2]^-$ and the meso-form $[M(R\text{-}asp)(S\text{-}asp)]^-$[13]. On the other hand we may assume that each aspartic acid molecule coordinates in a stereospecific manner in labile as well as in inert complexes[14]. The asymmetric carbon atom allows the coordination of aspartic acid as a tridentate ligand only in a way producing definite chirality of the coordination unity. Diastereoselectivity of a single ligand molecule has not yet been reviewed in an extensive manner. As this question is of great importance for the construction of a matrix-like model of definite structure, the following compilation will be limited to this field. Although investigations of mixed ligand complexes are quite numerous, and the number is increasing rapidly, the experimental material is still insufficient to draw general conclusions. Even though some examples of conformational analysis are known and predictions of selectivity can be made[15], the orientation of a secondary ligand in a chiral mixed ligand complex is still based on empirical data.

2.2. Stereoselective Coordination of a Single Ligand Molecule

2.2.1. Bidentate Ligands

The most thoroughly investigated bidentate ligands are certainly the 1,2-diamines. It was recognized quite early that the chelate rings of these ligands could not adopt a planar structure[16, 17]. The more stable gauche conformation may present λ or δ helicity so that substituents fixed on the carbon chain can adopt two non-equivalent positions. If the absolute configuration of such a ligand is determined, as for example in R(–)propylenediamine, one of the two conformations becomes more stable. On

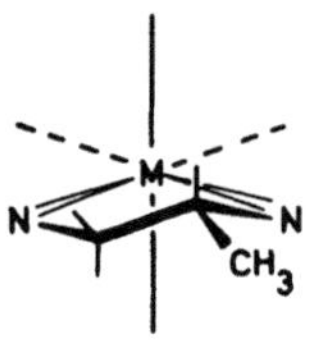

λ

1a

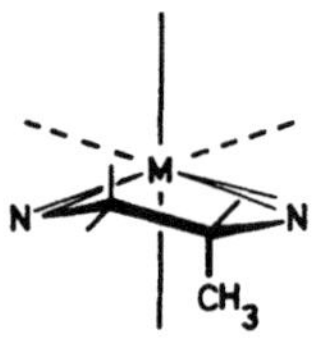

δ

1b

the base of conformational analysis the energy difference between *1a* and *1b* ranges between 0 to 3 kcal/mole[18]. This energy difference may be the main reason for the difference of the formation constants of metal complexes with racemic and meso-2,3-diaminobutane and their derivatives (Table 1).

In the racemic ligand both methyl groups are placed in an equatorial position *2a*. For the meso-compound this would be possible only in the unstable envelope conformation with the methyl groups in an eclipsed position *2c*. Such an envelope conformation was found for two Co^{III}-complexes with substituted tetramines[22, 23]. Recently Güntert and coworkers described the first examples of nickel(II)-tetramine complexes with a terminal chelate ring in envelope conformation[24]. In the gauche conformation one methyl group necessarily adopts an axial position *2b*. The assumption of an envelope conformation of the central diamine chelate ring of meso-DABTA could explain the large stability differences in alcalin-earth complexes, given in Table 1 for comparison, with respect to rac-DABTA. Whereas rac-DABTA can react as a hexadentate ligand in a similar way to EDTA, meso-DABTA would then be only

Table 1. Formation constants of some metal complexes with RS-diaminobutane (meso-DAB) and RR/SS-diaminobutane (rac-DAB) and the corresponding tetraacetic acids (DABTA)

	Meso-DAB[1])		Rac-DAB[1])		Meso-DABTA[2])	Rac-DABTA[2])
	$\log K_1$	$\log K_2$	$\log K_1$	$\log K_2$	$\log K$	$\log K$
Ni^{2+}	7.04	5.70	7.11	6.48		
Cu^{2+}	10.72	9.34	11.39	9.82		
Mg^{2+}					8.85	11.44
Ca^{2+}					9.67	12.34
Sr^{2+}					7.65	10.20
Ba^{2+}					6.53	8.52
H^+ pK_1	6.92		6.91		11.25	11.74
pK_2	9.97		10.00		6.27	6.12
pK_3					2.53	3.54
pK_4					1.80	2.41

1) From[19]) at 25°; 0,5 M KNO_3.
2) From[20, 21])at 20°; 0,1 M KCl.

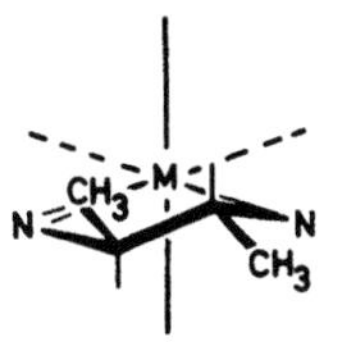

2a

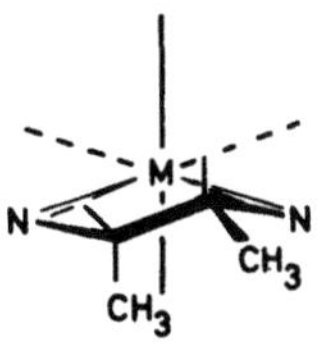

2b

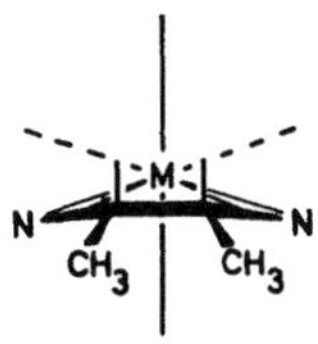

2c

pentadentate, one carboxylate group should point away from the coordination center.

There are practically no quantitative measurements concerning the equilibrium *1a*⇌*1b*. Dwyer and coworkers studied the correlation between chelate ring conformation and complex configuration in a series of ethylenediaminepropylenediamin-cobalt(III) complexes. From the results, they conclude that the methyl group of propylenediamine is always in an equatorial position[25, 26]).

NMR-measurements show that the diamine chelate ring of 1,2-(S)-diamino-propionic acid has λ-conformation in its planar Pt^{II}-complex[27]). The free carboxylate group takes an equatorial position *3*. For the nickel(II)-complex of the same ligand, Reilley and coworkers postulate, on the base of NMR-data, a change of configuration as a function of the pH-value of the solution[28]). At high pH-values, the carboxyl-

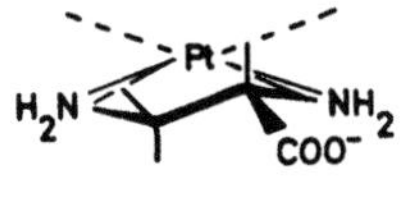

3

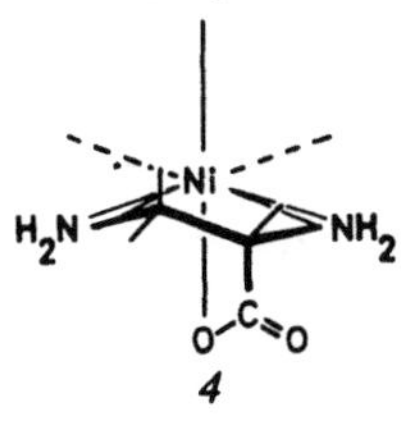

4

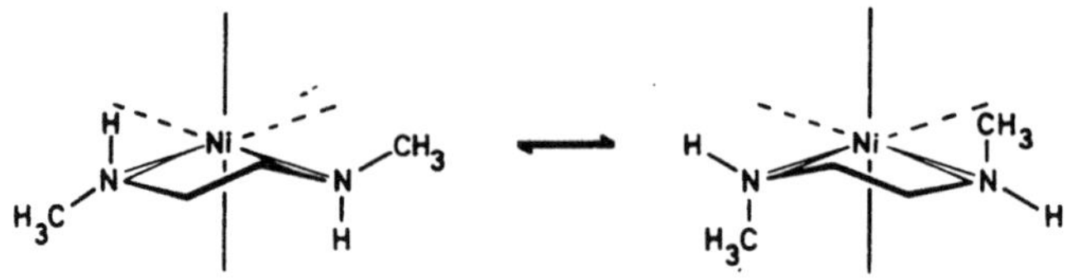

5a *5b*

ate group takes an axial position *4*. Only in this position with the diamine ring in δ-conformation can (S)-1,2-diaminopropionic acid act as a tridentate ligand. The same authors describe also a certain number of nickel(II)-complexes with N-substituted diamines[29]. These measurements indicate about 80% of *5a* in the conformational equilibrium.

For complexes with bidentate 1,2-diamines only one type of chelate ring conformation could be ascertained. The six-membered chelate rings of 1,3-diamines may exist in a chair *6a*, skew boat *6b* or boat *6c* conformation. The least stable of the three types is *6c* and was not taken into account so far. When the chelate ring carries

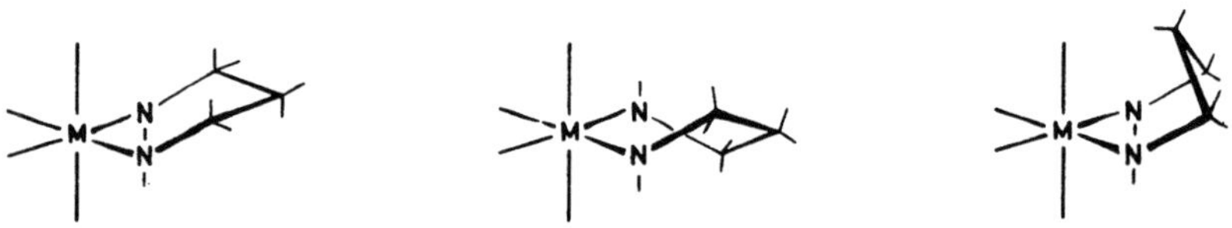

6a *6b* *6c*

no substituent, the chair conformation would be the most favorable. Conformational analysis predicts some flattening of the chelate ring due to the interaction of the axial hydrogen atoms of the α-carbon atoms with the ligands in the polar coordination site of the metal center[30]. This was confirmed by X-ray analysis for some complexes in the solid state[31, 32]. Information concerning the same phenomenon in solution is rather scarce. Appleton and Hall[33] carried out NMR-measurements for the first time with platinum and palladium complexes. In Pt^{II}-complexes they found a chair conformation not only for the meso- but also for the rac-2,4-diaminopentane. Contrary to this, the racemic ligand which, in a chair conformation, has one methyl group in an axial position, adopts a skew boat conformation in the Pt^{IV}-complex. This shows clearly the influence of the interaction between the axial substituent and the apical ligand in the octahedral Pt^{IV}-compound. In a similar manner the conformational change in $[Pt(l\text{-}CH_3\text{-}tn)(NH_3)_2]^{2+}$, as a function of temperature and in a solvent with high nucleophilic properties, *e.g.* $(CH_3)_2SO$, may be explained on the base of a pseudo-coordination of the solvent in the fifth and sixth coordination position of the square planar Pt^{II}-complex.

The first quantitative indications on conformational equilibria are given by Sarneski and Reilley[34] on the base of contact-shift measurements with nickel(II)-complexes. The results are summarized in Table 2.

In comparison to 1,2-diamines the five-membered chelate rings of amino-acids are much less puckered. Whereas dihedral angles of diamine chelate rings lie between

Table 2. Conformational equilibria of nickel(II) complexes with some 1,3-diamines

<table>
<tr><th></th><th>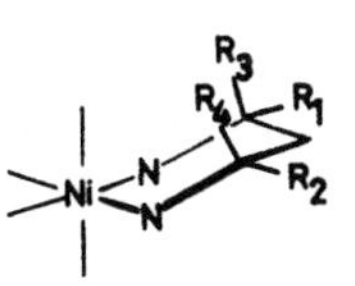</th><th>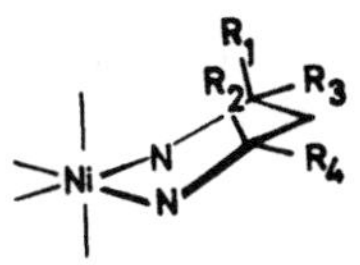</th><th>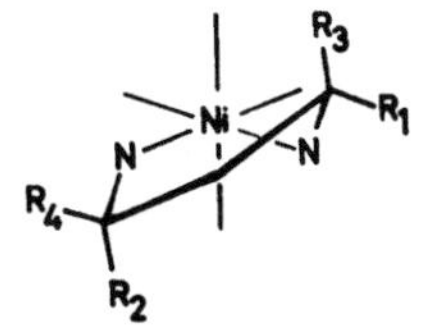</th></tr>
<tr><td>[Ni(meso-1,3-$(CH_3)_2$tn)]$^{2+}$
$R_1 = R_2 = -CH_3; R_3 = R_4 = -H$</td><td>100%</td><td>0%</td><td>0%</td></tr>
<tr><td>[Ni(rac-1,3-$(CH)_3)_2$tn)]$^{2+}$
$R_1 = R_4 = -CH_3; R_2 = R_3 = -H$</td><td colspan="2">54%</td><td>46%</td></tr>
<tr><td>[Ni(l-CH_3-tn)]$^{2+}$
$R_1 = -CH_3; R_2 = R_3 = R_4 = -H$</td><td>88%</td><td>4%</td><td>8%</td></tr>
<tr><td>[Ni(tn)]$^{2+}$
$R_1 = R_2 = R_3 = R_4 = -H$</td><td colspan="2">91%</td><td>9%</td></tr>
</table>

45° and 50°, structural analysis of amino-acid complexes in solid state shows dihedral angles varying from 0° to 30°[35]. Conformational analysis shows only small interactions between the substituent at the α-carbon atom and the ligand in the apical position of the complex. Therefore no substantial stability difference should exist between different conformations with the α-substituent in a pseudo-equatorial or pseudo-axial position[18]. Examples of amino acid complexes with substituents in axial positions are the solid *cis*-[Cu(L-ileu)$_2$][36], and *cis*-[Cu(D-ala)$_2$][37]. It must nevertheless be mentioned that in solids the requirement for lattice packing may determine the molecular conformation. Results obtained from solids can only be valid in a restricted sense for the situation in solution.

Wellmann and coworkers give some interesting features for aminoacid complexes in solution[38]. They conclude from the quasi-enantiomeric shape of the CD-spectra of the Cu^{2+}-complexes with L-histidine and L-l-CH_3-histidine that the equilibrium state of the system *7a*⇌*7b* is strongly in favor of *7a* in the case of bidentate amino-

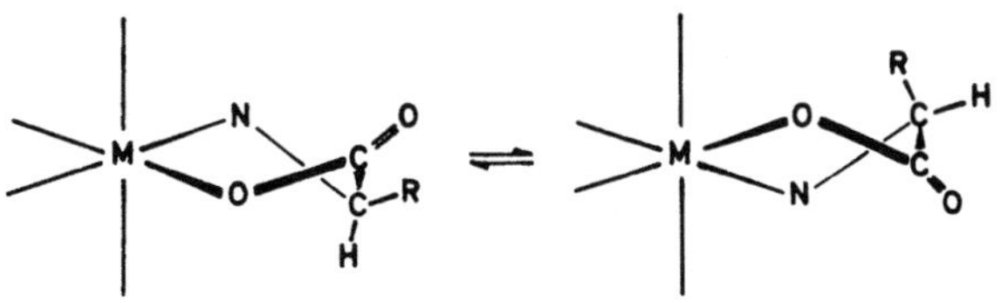

7a *7b*

acids with voluminous groups R, for example L-l-CH_3-histidine or L-phenylalanine. L-histidine or L-3-CH_3-histidine on the other hand can act as tridentate ligands; the carboxylate group is then forced into an axial position and the complex must adopt the structure corresponding to *7b*. The case is somewhat similar to that of the Ni^{2+}-complex of L-diaminopropionic acid mentioned earlier. When aspartic or glutamic acid react as bidentate ligands, for example in Pt^{II}-complexes, NMR-measurements show the noncoordinated rest of the ligand in equatorial position[27]. For the N-methyl group in planar $[Pt^{II}(sar)Cl_2]^-$ a pseudo-axial position was proposed[27], whereas it was found to be in an equatorial one in the octahedral $[Co(en)_2(sar)]^{2+}$[39].

Only little is known about the structure of six-membered chelate rings in complexes with β-aminoacids. The most probable is a distorted boat conformation with the more voluminous substituent neighboring the amino group in an equatorial arrangement.

2.2.2. Tridentate Ligands

For tridentate ligands we may distinguish two structurally different types: linear *A* and branched *B* ones. In linear ligands *A* one of the ligand atoms is necessarily a

(A) (B)

member of the chain. This ligand atom becomes a new center of chirality for unsymmetrical molecules and for complexes with a nonplanar structure of the chelate rings. Linear ligands may be coordinated in a peripheric as well as in a facial manner. For branched ligands *B* only a facial coordination is possible. This is an important restriction for the determination of the geometry of the ligand sphere of a coordination compound. When the point of chain branching is at the same time a center of chirality – which will be the case in most examples – the coordination of such ligands will be stereospecific. Such a property will therefore limit the number of isomers for bis- and mixed-ligand complexes. Some of the most important optically active tridentate ligands are given below:

A: *linear ligands:*

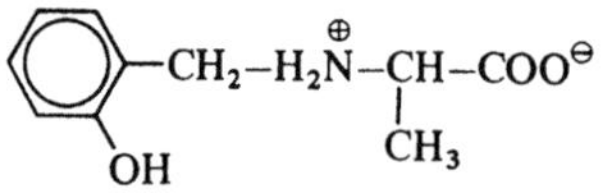

8

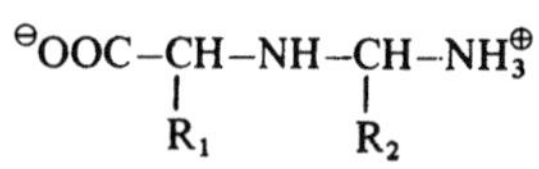

9

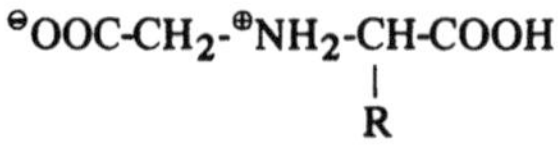

10

B: *branched ligands:*

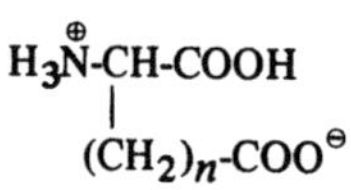

11 $n = 1,2$

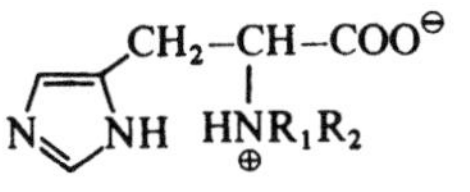

12

a) $R_1 = R_2 = -H$
b) $R_1 = -H; R_2 = -CH_3$
c) $R_1 = -H; R_2 = -CH_2-$ 
d) $R_1 = R_2 = -CH_3$

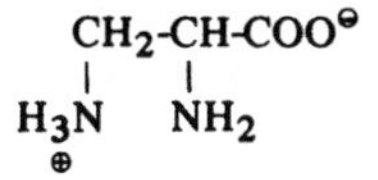

13

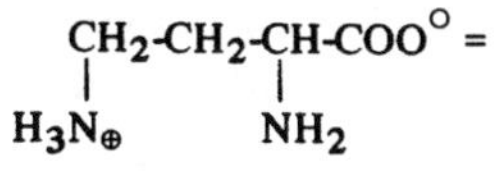

14

The diastereospecificity of branched tridentate ligands has been confirmed for all examples studied so far. For the system $[Co(L\text{-hist})_2]^+$ L. J. Zompa[40] found only three of the five theoretically possible isomers. Freeman and Liu[41] obtained the same result for the Co^{III}-complexes with the optically active diaminopropionic acid *13*. The racemic compound on the other hand gave the five isomers expected. The optically active complexes show the absolute configuration corresponding to the structure of the ligand. The corresponding diastereoisomers with the opposite absolute configuration do not exist by reason of the extremely high tension of the asymmetric carbon atom. The structure of one of these isomers has been established by X-ray diffraction analysis[42], the attribution of the others is then made by comparison of spectral and chiroptical properties. The complexes of 2,4-diaminobutyric acid *14* show the same properties[43]. Introduction of a six-membered chelate ring has only little influence on chiroptical properties. Therefore a boat conformation of the six-membered chelate ring is not very probable, although NMR-data may be interpreted on the base of an equilibrium between a chair and a boat conformation. Legg and Cook showed the stereospecific coordination of a single molecule of aspartic acid[44]. From the six isomers theoretically possible, only three, *15a, 15b* and *15c*, were found.

In this case too, the absolute configuration of the isomers correspond to those

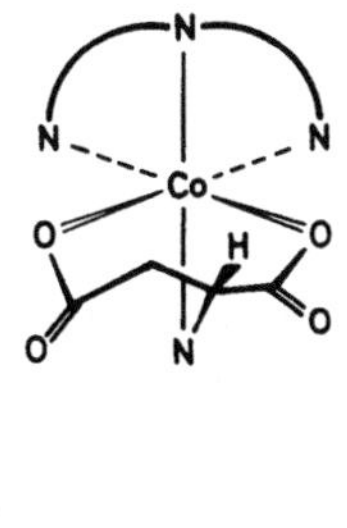

15a
s-*cis*

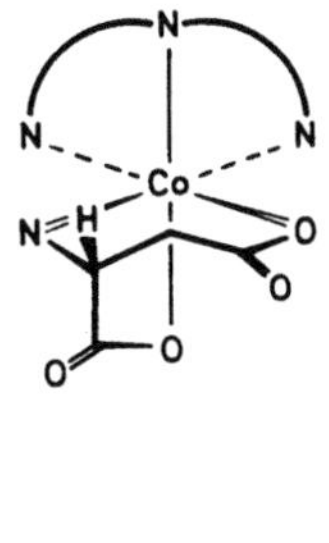

15b
Δ (C_2)-u^2-*cis*

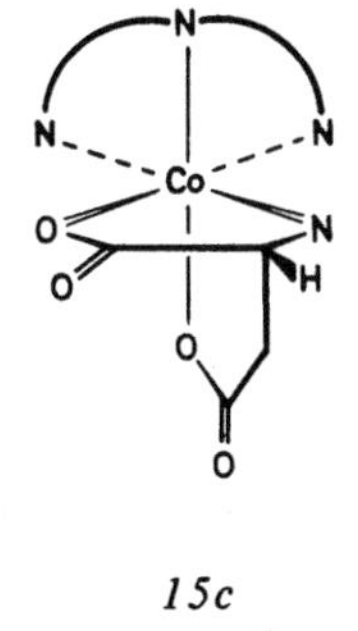

15c
Λ (C_2)-u^1-*cis*

expected from the R-configuration of the asymmetric carbon atom of the ligand.

Information for linear tridentate ligands is rather rare. Such information is very often of indirect nature, in a way that stereoselectivity in the formation of bis- or mixed-ligand complexes may be indicating a stereoselective coordination of the single ligand. Ritsma found that in bis-complexes of Ni^{2+} and Co^{2+} with α-N-hydroxybenzylamine *8* the racemic (M-LL/M-DD) complexes are more stable than the meso-complex (M-DL)[12]. Although stereoselectivity is sometimes inversed and the meso-compounds more stable as the racemic ones, the stability differences are of comparable magnitude, as those found for the bis-complexes of the histidine derivatives 12a–c. $\Delta\Delta G$-values between 0,5-1 kcal/mole were found by Angelici and coworkers for mixed ligand complexes of Cu^{2+}-L-valinemonoacetate with several D- and L-amino acids. The stability constants were higher for the L-amino acids.

We may assume that these results are due to a preferential orientation of both ligands. It is not known if this orientation is already present in each ligand – different or identical in nature – or if it is a result of the interaction between both ligands. It is shown by the studies of complexes of dipeptides that both factors may be more or less responsible for diastereoselectivity in bis- or mixed ligand complexes.

The absolute configuration of [Co(glygly)]$^-$ is known[45]. The complex has C_2-symmetry and the antipodes *16a* and *16b* have been separated by chromatography on starch[46]. In the coordination octahedron the tridentate dipeptide ligand takes a

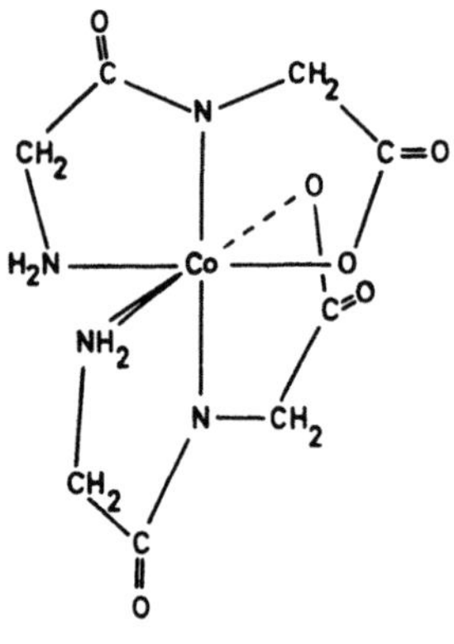

(+) [Co(glygly)]$^-$

16a

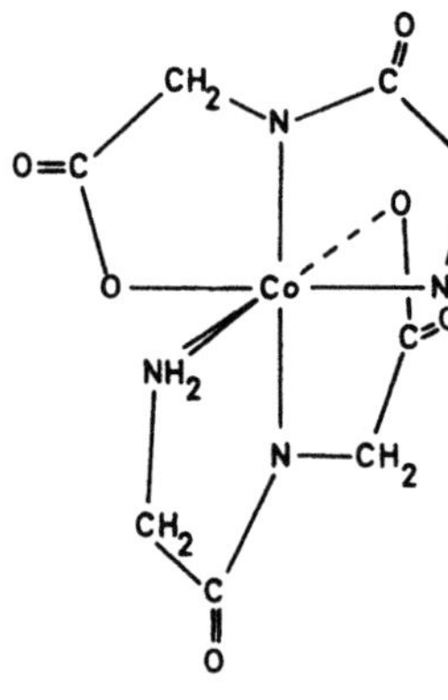

(–) [Co(glygly)]$^-$

16b

helicoidal structure[47], in such a way that two enantiomeric conformations are possible. In complexes with dipeptides from optically active amino acids, these two conformations become diastereoisomers as shown by *17a* and *17b*. The complex formation seems to be selective for [Co(dipep)$_2$]$^-$-complexes (dipep = L-ala-gly, gly-L-leu, and L-ala-L-ala), only the form corresponding to (–) [Co(glygly)$_2$] could be found found[46, 48]. The position of the asymmetric center, be it at the O- or at the N-ter-

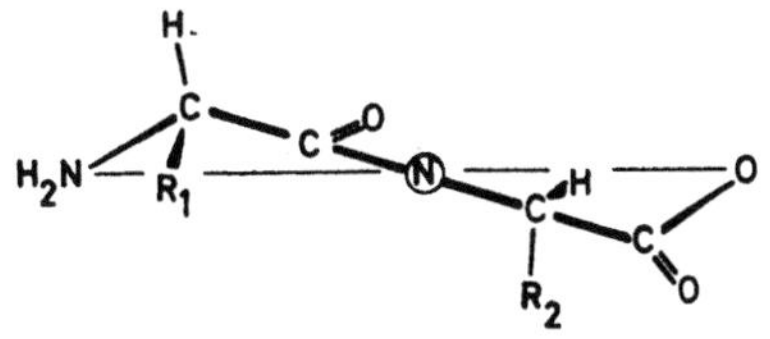

17a

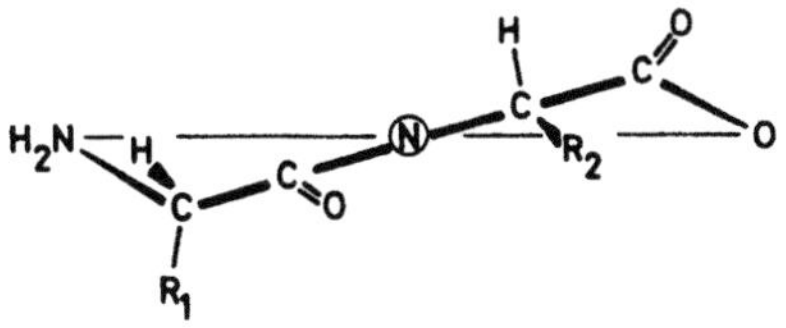

17b

minal aminoacid, seems to be of no importance for the stereoselectivity of the complex. An analogous selectivity is presumed for the different geometrical isomers of bis-complexes with glycyl-L-histidin[49]. On the base of chiroptical properties, Michailidis and Martin assume a ligand conformation of definite chirality for the corresponding complexes with L-hist-gly, gly-L-ala and L-ala-gly[50]. In contrast to these results the chiroptical properties of mixed amine-dipeptide complexes, containing only one dipeptide molecule, for example $[Co(dien)(dipep)]^+$ and $[Co(NH_3)_3(dipep)]^+$ (dipep = glygly, L-ala-gly, L-ala-L-ala and gly-L-ala), may be explained on the base of a sole vicinal effect[51]. Dipeptides from two optically active amino acids of the same absolute configuration (*e.g.* S-configuration as in the examples *17a* and *17b*) one substituent is always in a pseudo-axial and the other in a pseudo-equatorial position independently of the chirality of the conformation of the ligand. Therefore equilibrium constants of the formation of labile complexes with dipeptides from amino acids with opposed configuration should be higher than those for the corresponding diastereoisomers. This is confirmed by data given in Table 3 with the exception of leucyl-phenylalanine. The number of data is of course too small to allow general conclusions. Further investigations in this field seem useful not only concerning structural properties of inert Cobalt(III)-complexes but also equilibrium data of labile complexes with metal ions of biological interest as Cu^{2+}, Zn^{2+}, Mn^{2+} and others.

Table 3. Stability constants of metal-complexes with some diastereoisomeric dipeptides

$(AA)_1$	$(AA)_2$	Co^{2+}		Ni^{2+}		Cu^{2+}		Zn^{2+}		Ref.
$(H_2N\text{-})$	(HOOC-)	$\log K_1$	$\log K_2$	$\log K_1$	$\log K_2$	$\log K_1$	$\log K_2$	$\log K_1$	$\log K_2$	
L-ala	L-ala	2,63								1)
L-ala	D-ala	2,83								1)
L-leu	L-tyr	2,42	4,48	3,23	5,99			2,98	5,66	1)
D-leu	L-tyr	2,81	5,07	3,73	6,66			3,39	6,24	1)
L-leu	L-phen			2,85		3,82				2)
L-leu	D-phen			1,7		3,1				2)

1) Li, N. C., Miller, G. W., Solony, N., and Gillis, B. T.: J. Amer. Chem. Soc. *82* 3737 (1960).
2) Karczinsky, F., and Kupriszewski, G.: Rocz. Chem. *41* 1019 (1967).

2.2.3. Tetradentate Ligands

With tetradentate ligands we can distinguish four basic structural types:

C *D* *E* *F*

Whereas for ligands of type *C* and *F* there are three, alternatively two, possible topographies in octahedral complexes, only one possible topography exists for ligands of type *D* and *E*. Nevertheless geometrical isomers may exist for octahedral complexes with ligands of type *D* when the nature of the two terminal groups fixed on the branching point is nonidentical. Geometrical isomers of type *E* need at least one terminal group to be different from the two others. The stability of the isomers of all these types of ligands is determined by the size and the conformation of the chelate rings, the position of substituents and the nature of the ligand atoms as well as the metal center.

2.2.3.1. Linear Ligands. Most of the investigations were accomplished with linear tetramine ligands, among which triethylenetetramine (2,2,2-tet) and its substituted derivatives are well known[52]. By coordination of such ligands to a metal ion, the two secondary nitrogen atoms become two new centers of chirality. Five geometrical arrangements are then possible *18a–e*. Except for the meso-*trans* isomer *18e*, all

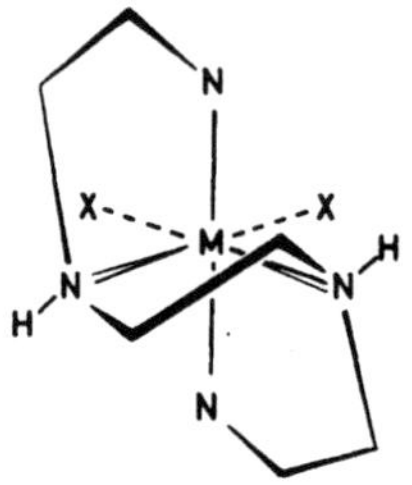

Λ-*cis*-α(RR)
18a

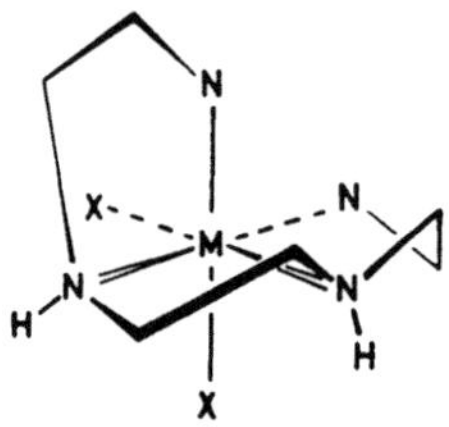

Δ-*cis*-β(RS)
18b

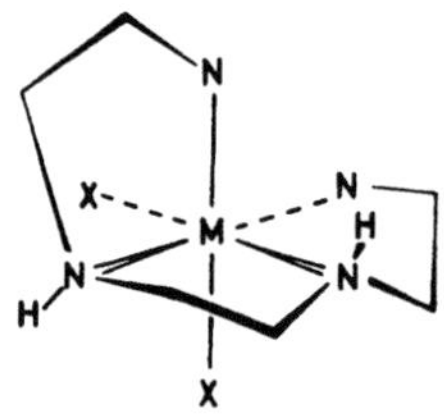

Δ-*cis*-β(RR)
18c

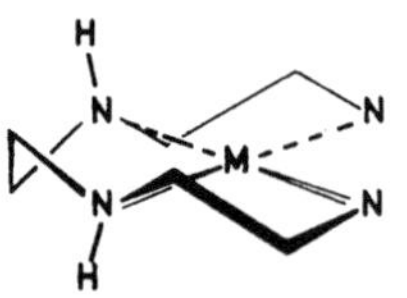

trans (SS)
18d

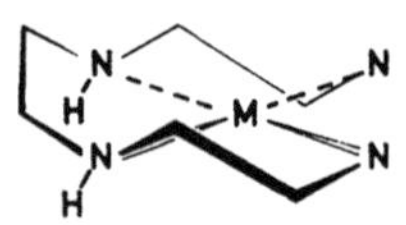

trans (RS)
18e

the forms could be obtained as inert cobalt(III)-complexes with the unsubstituted ligand (2,2,2-tet)[53]. According to studies of Sargeson and Searle[54] the three *cis*-isomers are more stable than the *trans* compounds. The conformation of chelate rings lying in the equatorial plane is given by the configuration of the secondary nitrogen atoms. A chelate ring joining an equatorial and a polar coordination site seems sufficiently flexible to make λ and δ conformation possible, independently of the configuration of the nitrogen atoms[53, 55]. The observation that the *cis*-α isomer with (3S, 8S)-$(CH_3)_2$-2,2,2-tet always shows Λ-configuration[56, 57] indicates that a δ-conformation of the two terminal chelate rings (equatorial positions of the methyl groups, see Table 4) favors a Λ-configuration of this isomer.

At the present time about a dozen optically active derivatives with the basic framework of 2,2,2-tet are known. All of them react highly stereospecifically in such a way as to place the substituents mostly in an equatorial position. From these results Saburi and Yoshikawa conclude[22] that a suitable choice of substituents will allow fixing of the conformation of the chelate rings and the absolute configuration of a given geometric isomer. They further predict that it should be possible, in a similar way, to obtain isomers which for the unsubstituted tetramines either cannot be synthesized or only with great difficulties.

Finally Brubaker and Euler conclude from a conformational analysis that the stability difference between Δ-*cis*-β and Λ-*cis*-β-oxalato- N,N′-bis(2-picolyl)1,2-(S)-diaminopropanecobalt(III) (Λ-*cis*-β- $[Co(pnpic)C_2O_4]^+$) must be at least 3 kcal/mole[58].

Table 4 gives the different structural elements of the synthesized isomers of Co^{III}-complexes with optically active tetramines of the type 2,2,2-tet known so far. Only a small number of optically active tetramines of the type of 2,3,2-tet *19* and 3,2,3-tet *20* forming complexes with six-membered chelate rings are known. Despite the great preference for a trans-topography of these ligands, *cis*-β-complexes can be obtained under certain conditions. The simplest way to obtain compounds with such a structure is to synthesize mixed ligand complexes with a bidentate secondary ligand. Harrington and coworkers[67] describe Co^{III}-complexes of 4,8-diaza-2,10-undecanediamine, a ligand with a basic structure of type *19* and substituents in the two terminal chelate rings. The authors studied the optically active (racemic) as well as the meso-ligand. Both diastereomeric ligands react in a stereospecific way. Only the three compounds *21a, 21b,* and *21c* were obtained.

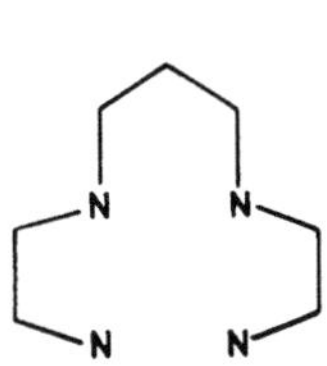

19

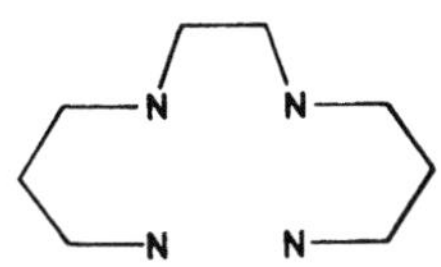

20

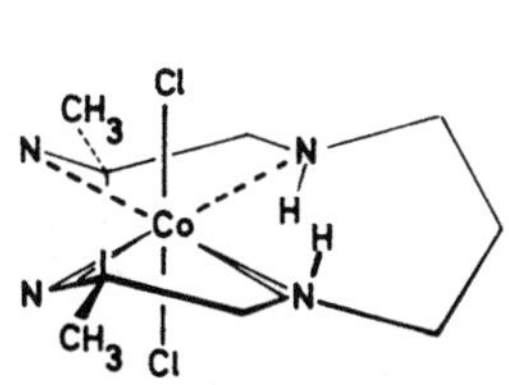

trans-$[Co((SS)\text{-}Me_2\text{-}2,3,2\text{-tet})Cl_2]^+$

21a

Table 4. Structures of Co^{III}-complexes with substituted tetramines of the type of 2,2,2-tet

Ligand	Complex-configuration[2])	N_1	N_4[3])	N_7[3])	N_{10}	C_2	C_3	C_5	C_6	C_8	C_9[4])	Conformation of chelate rings N_1N_4	N_4N_7	N_7N_{10}	Ref.
(5R,6S)Me_2-trien	meso-*trans*	–	S	R	–	–	–	R(e)	S(e)	–	–	δ	ε	λ	23)
(5R,6S)Me_2-trien	rac-*trans*	–	R	R	–	–	–	R(a)	S(e)	–	–	λ	δ	λ	23)
(5R,6S)Me_2-trien	*Δ-cis-β*[5])	–	S	R[1])	–	–	–	R(e)	S(a)	–	–	δ	λ	δ	23)
(2S,5R,6S,9S)Me_4-trien	rac-*trans*	–	S	S	–	S(e)	–	R(e)	S(a)	–	S(a)	δ	λ	δ	23)
(2S,5R,6S,9S)Me_4-trien	*Δ-cis-β*	–	S	R	–	S(e)	–	R(e)	S(a)	–	S(e)	δ	λ	δ	23)
(3S,9R)Me_2-trien	*Λ-cis-β*	–	S	S[1])	–	–	S(e)	–	–	R(e)	–	δ	λ	λ	22)
(3S,8R)Me_2-trien	meso-*trans*	–	S	R	–	–	S(e)	–	–	R(e)	–	δ	ε	λ	22)
(3S,8S)Me_2-trien	*Δ-cis-β*	–	S	R[1])	–	–	S(e)	–	–	S(e)	–	δ	λ	δ	59)
(3S,8S)Me_2-trien	*Λ-cis-β*	–	R	R	–	–	S(e)	–	–	S(e)	–	δ	λ	δ	56, 57)
(3S,8S)Me_2-trien	rac-*trans*	–	S	S	–	–	S(e)	–	–	S(e)	–	δ	λ	δ	60)
(2S,9S)Me_2-trien	*Δ-cis-β*	–	S	R[1])	–	S(e)	–	–	–	–	S(e)	δ	λ	δ	59)
(2S,9S)Me_2-trien	*Λ-cis-α*	–	R	R	–	S(e)	–	–	–	–	S(e)	δ	λ	δ	59)
(2S,5R,9S)Me_3-trien	*Δ-cis-β*	–	S	R[1])	–	S(e)	–	R(e)	–	–	S(e)	δ	λ	δ	59)
(2S,5R,9S)Me_3-trien	*Λ-cis-α*	–	R	R	–	S(e)	–	R(e)	–	–	S(e)	δ	λ	δ	59)
(2S,5S,9S)Me_3-trien	*Δ-cis-β*	–	S	R[1])	–	S(e)	–	S(a)	–	–	S(e)	δ	λ	δ	59)
(5R)-Me-trien	*Λ-cis-α*	–	S	S[1])	–	–	–	R(e)	–	–	–	δ	λ	λ	59)
(5R)-Me-trien	*Δ-cis-β*	–	S	R[1])	–	–	–	R(e)	–	–	–	δ	λ	δ	55)
(1,5R,10)Me_3-trien	rac-*trans*	R	S	S	R	–	–	R(e)	–	–	–	δ	λ	δ	60)
(1 5R,10)Me_3-trien	*cis*[6])	–	–	–	–	–	–	–	–	–	–	–	–	–	60)
(1,2S,9S,10)Me_4-trien	rac-*trans*	R	S	S	R	S(e)	–	–	–	–	S(e)	δ	λ	δ	60)
(1,3S,8S,10)Me_4-trien	rac-*trans*	R	S	S	R	–	S(e)	–	–	S(e)	–	δ	λ	δ	60)
(5S,6S)Me_2-trien	*Δ-cis-β*	–	R	R	–	–	–	S(e)	S(e)	–	–	λ	δ	δ	55)
R,R-trans-N,N'-bis(β-amino-ethylcyclohexane-diamine	*Λ-cis-β*[7])	–	S	S	–	–	–	R(e)	R(e)	–	–	δ	λ	λ	55)

Ligand	Complex-configuration²)	N_1	$N_4$³)	$N_7$³)	N_{10}	C_2	C_3	C_5	C_6	C_8	$C_9$⁴)	Conformation of chelate rings N_1N_4	N_4N_7	N_7N_{10}	Ref.
N,N'-bis(2-picolyl)-(1S)-methyl-1,2-diaminoaethane	Δ-*cis*-β⁷)	–	R	R	–	–	–	S(e)	–	–	–	–	δ	–	58, 61, 62)
N,N-bis(2-picolyl)-(1S, 2S)-dimethyl-1,2-diaminoaethane	Δ-*cis*-β⁷)	–	R	R	–	–	–	S(e)	S(e)	–	–	–	δ	–	61)
R,R-trans-N,N'-bis(2-picolyl)-cyclohexanediamine	Λ-*cis*-β⁷)	–	S	S	–	–	–	R(e)	R(e)	–	–	–	λ	–	63)
N,N'-bis((2S)-amino-3-phenylpropyl) trans-R,R-diaminocyclohexane	Δ-*cis*-β⁷)	–	S	R¹)	–	S(e)	–	R(e)	R(e)	–	S(e)	δ	λ	δ	64)
	rac-*trans*⁷)	–	S	S	–	S(e)	–	R(e)	R(e)	–	S(e)	δ	λ	δ	64)

1) Indicates the angular nitrogen atom.
2) The absolute configuration about the central metal ion is indicated as Λ or Δ according to the tentative rule of[65].
3) The configuration of secondary nitrogen atoms is represented as R and S according to the rules suggested by[66] neglecting the substituents in the trien framework.
4) (e) alternatively (a) indicate equatorial respectively axial position of the substituent.
5) The indicated structure stands for the compound isolated from the less soluble diastereoisomer with α-brom-α-camphersulfonic acid.
6) The compound is mentioned but without any additional information.
7) For systematic reasons the atoms forming the central chelate ring are placed in the columns N_4-C_5-C_6-N_7.

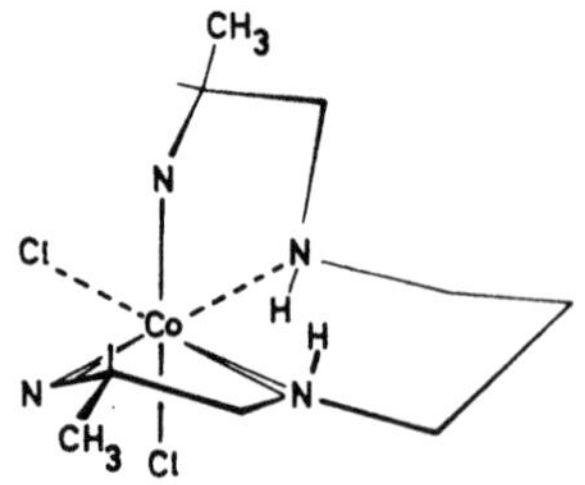

Λ-*cis*-β-[Co((SS)-Me_2-2,3,2-tet)Cl_2]$^+$
21b

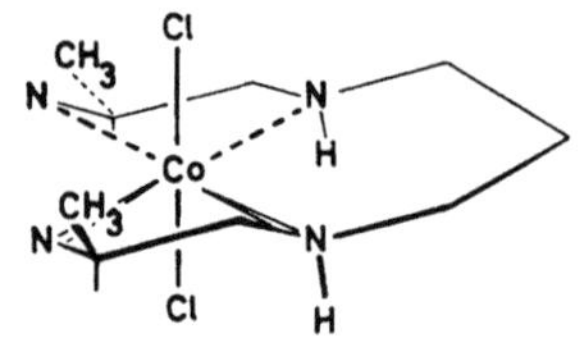

trans-[CO((RS)-Me_2-2,3,2-tet)Cl_2]$^+$
21c

Once more, we observe that the equatorial position of the substituents is the structurally determining factor. We may further conclude from these examples that it will be easier for a six-membered chelate ring to adopt a skewboat conformation than for a five-membered chelate ring to adopt an envelope conformation. On the other hand, it has been shown by the examples mentioned earlier[24] (the square planar nickel(II)-complexes with (2R, 7S, 10R)-2,10-diamino-5,5,7-trimethyl-4,8-diazaundecane and the corresponding (2R, 7R, 10R)-diastereoisomer), that the preference for the different conformations of five and six-membered chelate rings depends on the number and the position of the substituents. The great similarity of NMR-, ORD- and CD- spectra lead to the conclusion that the two diastereomeric complexes are chromophores of identical chirality concerning the framework of the ligand molecule. This can only be realized in the structures *22a* and *22b*. In these

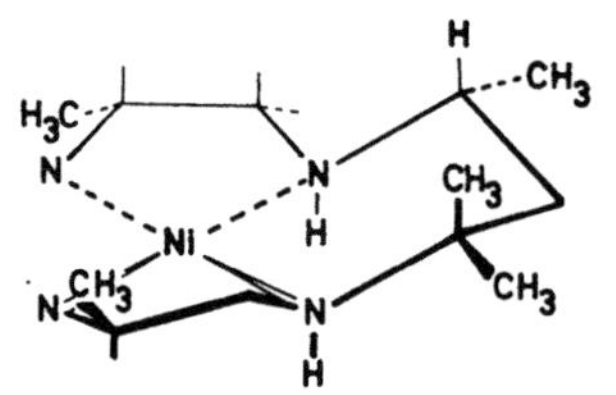

22a

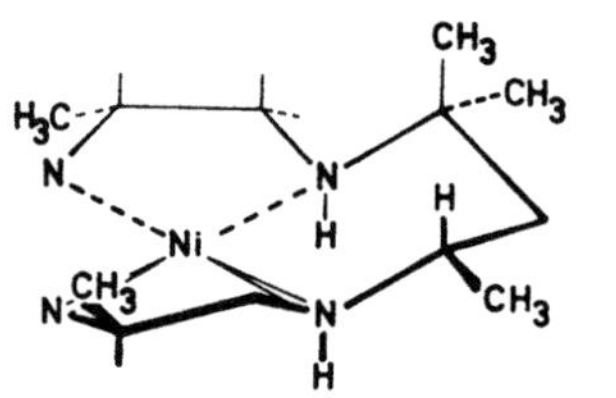

22b

compounds the six-membered ring prefers – contrary to what was observed in *21a* – in both cases the more stable chair conformation, whereas one of the two five-membered rings shows the unstable envelope conformation. The structure of *22a* has been confirmed by X-ray diffraction analysis[68]. Diastereoisomerism of the complexes *22a* and *23b* is given by the interchange of C_5 and C_7 relative to the different conformations of the terminal chelate rings.

It is well known that for complexes with the 2,2,2-tet-type ligands an envelope conformation is not found[53], unless imposed by the position of substituents[22, 23] (see Table 4). Nevertheless such a conformation is very often found in macrocyclic complexes with ligands of the type "tetraazacyclotetradecanes"[69].

An example of a ligand containing a single substituent in the six-membered ring has been described by DeHayes and coworkers[70]. With (5S)-Me-2,3,2-tet they could isolate two Co^{III}-complexes for which they proposed the structure *23a* and *23b*.

The proposition *23a* is surprising. Beside the skew-boat conformation of the central chelate ring determined by the RR[a])-conformation of the secondary nitrogen atoms, the substituent is placed in a axial position. An equatorial position could be obtained

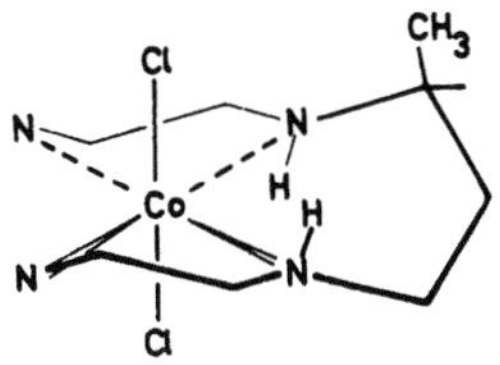

trans-RR-[Co((5S)Me-2,3,2-tet)Cl_2]$^+$

23a

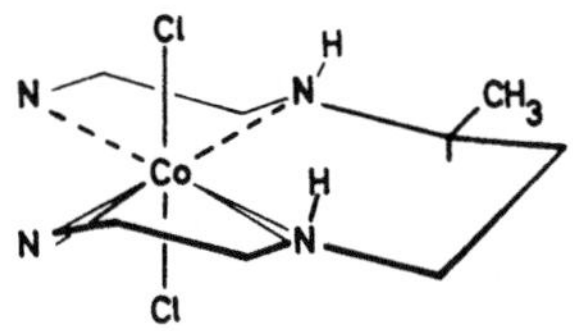

trans-RS-[Co((5S)Me-2,3,2-tet)Cl_2]$^+$

23b

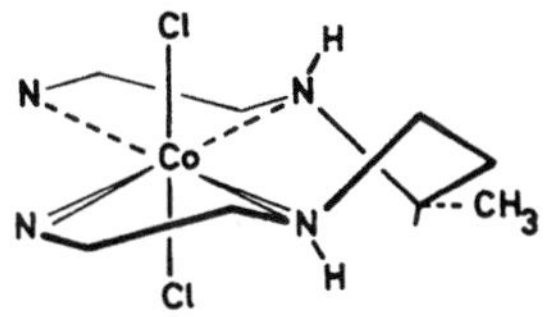

23c

for this substituent conserving the relative conformations of the three chelate rings by an inversion of both secondary nitrogen atoms giving the SS[a]) configuration as shown by *23c.*

The same authors describe also (6R)-Me-3,2,3-tet, the only optically active ligand known so far with the basic structure *20*[70, 71]. This ligand too reacts stereospecifically. Beside the cis-isomer with Λ-*cis*-β-structure, the *trans*-SS-[Co(6R)-Me-3,2,3-tet)Cl_2]$^+$ is the only one of the six possible *trans* isomers which has been found. This result corresponds to the conformational analysis which predicts *24* as the most stable one. As already mentioned, it is possible to fix the absolute configuration of a given isomer by a suitable choice of substituents. A selection between different geometrical isomers – *e.g.* on the base of chelate ring size – seems to be less easy. It

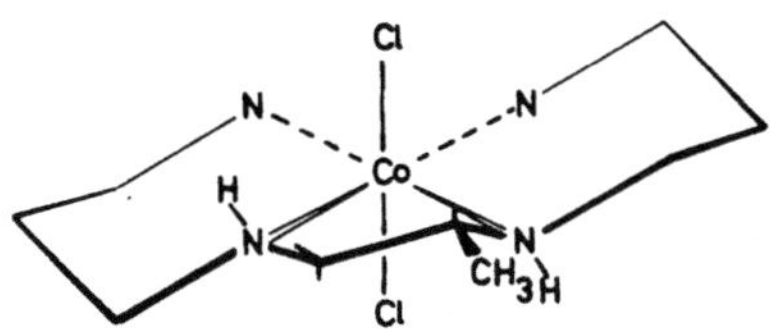

trans-SS-[Co((6R)-Me-3,2,3-tet)Cl_2]$^+$

24

would therefore be difficult – especially for labile compounds – to determine in advance the geometry and the chirality of such a system on the base of substituents only. However, such a determination can be realized by choosing an adequate combination of coordinating atoms. Two other sequences of donor atoms in liner tetra-

[a]) The designation R and S of the secondary N-atoms is chosen for the unsubstituted ligand molecule and is not changed after the introduction of substituents.

dentate ligands have been studied: ONNO in ethylenediaminediacetic acid (EDDA), and its derivatives and the compounds of the type of dithiadiamines (NSSN).

When Co^{III}-EDDA-complexes are synthesized under conditions favoring equilibrium state, the *cis-α*-compound is essentially obtained[72]. *Cis-β*-complexes can also be obtained, most favorably as mixed ligand complexes with a bidentate ligand, *e.g.* carbonate, oxalate or 1,2-diamines[73–75]. But in all cases known, it is the *cis-α*-isomer which is formed in a greater amount. Reilley and co-workers conclude from the simplicity of the NMR-contact shift spectra of $[Ni(EDDA)(H_2O)_2]$ that the compound is present in only one isomeric form, and assume that this is the *cis-α*-form[28].

Of great interest in this and related aminopolycarboxylate-complexes is the possibility of differentiating the two hydrogen atoms situated at the α-carbon of glycinate rests[76]. Exchange reactions of these atoms are stereoselective and significantly faster for the hydrogen atoms in the "out of plane" – position (H_A) than for those in the "in plane" – position (H_B).

Replacement of these hydrogen atoms by methyl groups introducing two asymmetric centers on the other hand produces only small stereoselectivity concerning the absolute configuration of the complex.

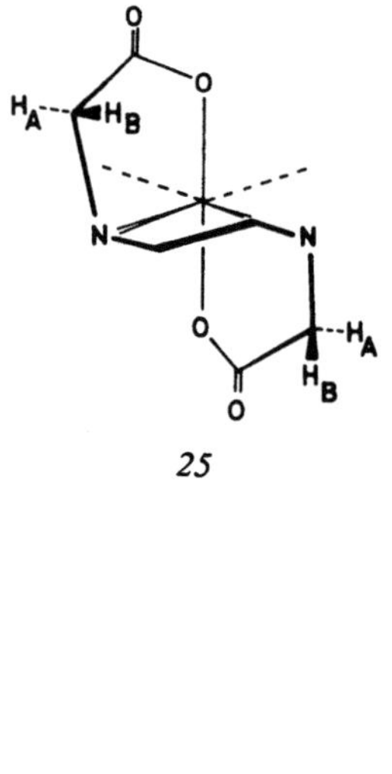

25

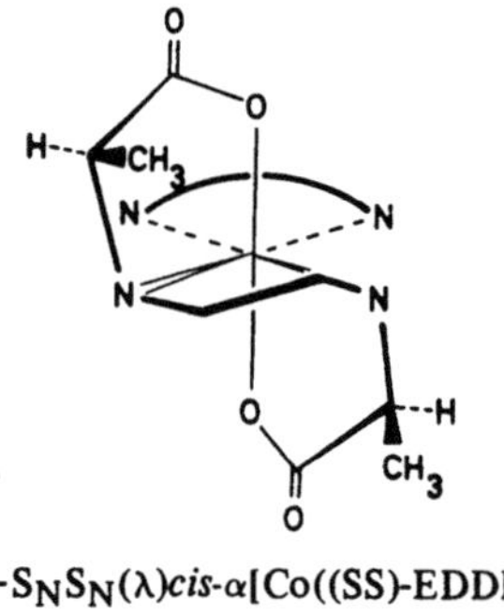

endo-Λ-S_NS_N(λ)*cis*-α[Co((SS)-EDDP)en]$^+$

26a

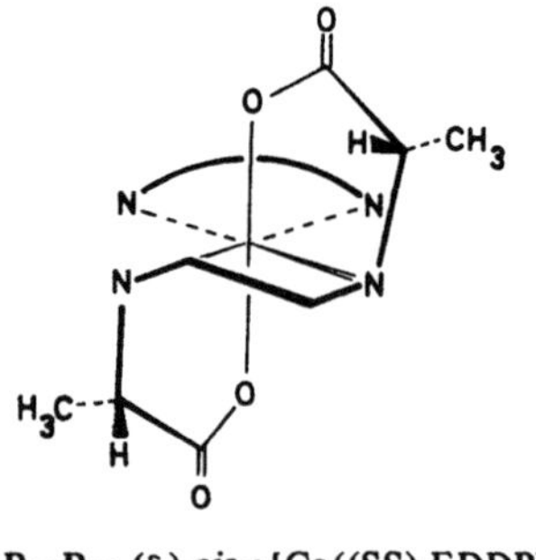

exo-Δ-R_NR_N-(δ)-*cis*-α[Co((SS)-EDDP)en]$^+$

26b

[Co((S,S)-EDDP)(en)]$^+$, a complex with the optically active ethylenediamine-di-2-propionic acid, gives nearly equal amounts of Λ-*cis*-α *26a* and Δ-*cis*-α *26b*[77]. A small quantity of Δ-*cis*-β was also obtained. When in the same complex, ethylenediamine is replaced by optically active R-propylenediamine the Δ-*cis*-α-compound is formed in a considerably greater amount with respect to the Λ-*cis*-α. The same result is obtained (more Δ-*cis*-α than Λ-*cis*-α) when the reaction is performed with R,R-EDDP instead of S,S-EDDP. This observation shows clearly that in this case the diastereoselectivity is due to the presence of the optically active bidentate diamine and not to the two asymmetric centers in the amino acid groups of the tetradentate ligand. This is another example where diastereoselectivity of a multidentate optically active ligand becomes important only by the introduction of a second optically active ligand into the coordination entity (cf. p. 12). It seems therefore that the introduction of substituents in the terminal chelate rings of mono- or diamino-polycarboxylates is not sufficient to determine the absolute configuration of a 1 : 1-complex.

A higher selectivity could be expected from a ligand with the EDDA-framework and a substituent in the central chelate ring. If this substituent takes an equatorial position it will determine the conformation of the central chelate ring and hence the configuration of both secondary nitrogen atoms in the *cis*-α isomer (S,S for λ conformation). Only one absolute configuration of the complex should be realized. Unfortunately, ligands of this type are not known so far, but the assumption becomes probable by the examples known for the analogous tetradentate dithiadiamines. Compounds with the basic structure of 1,8-diamino-3,6-dithiaoctane *27* forming complexes with three five-membered chelate rings are found exclusively as the *cis*-α-isomer in their complexes[78–81]. Introduction of six-membered chelate rings leads to the other geometrical isomers too. When the six-membered ring lies in the center, as is the case for the compound *28*, *cis*-β and *trans*-isomers are obtained; with the ligand *29* the isolated compounds have *cis*-α or *cis*-β-structure[78]. Optically active

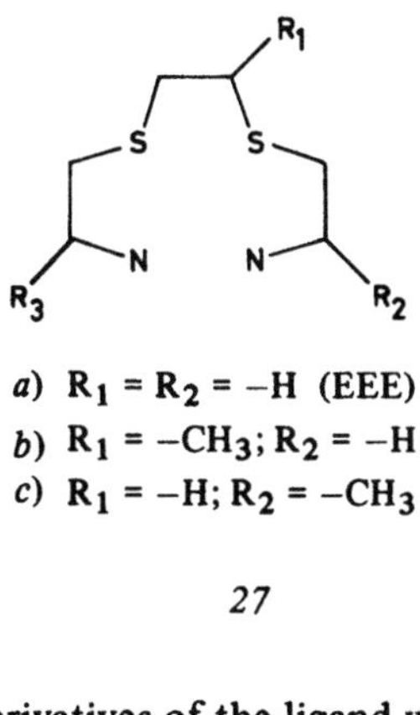

a) $R_1 = R_2 = -H$ (EEE)
b) $R_1 = -CH_3; R_2 = -H$ (EPE)
c) $R_1 = -H; R_2 = -CH_3$ (PEP)

27

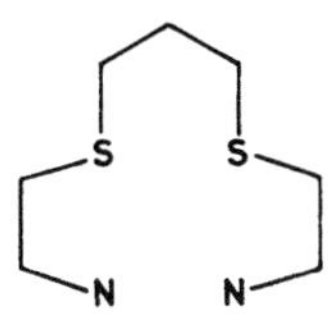

28 (ETE)

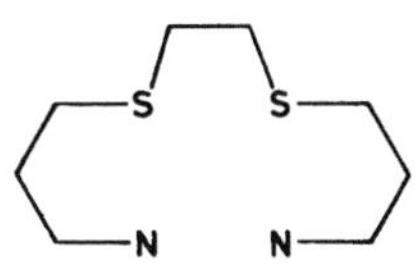

29 (TET)

derivatives of the ligand with basic structure *27* have been prepared. *27b* shows an asymmetric center in the S-S- and *27c* in the two terminal S-N-chelate rings. We may assume that a S-N-chelate ring is considerably more puckered than the N-O-chelate rings in EDDA-complexes. Despite this the stereoselectivity of optically active *27c* is rather small as was the case with EDDP. Bosnich and Philipp[82] found for the Co^{III}-complexes of (S,S)-*27c* an equilibrium mixture containing exo-Λ-*cis*-α-$[Co((S,S)\text{-}(27c))X_2]^+$ and the diastereomeric endo-Δ-*cis*-α-$[Co((S,S)\text{-}27c))X_2]^+$, ($X = Cl^-, SCN^-, NO_2^-$). The Λ/Δ-ratio was about 2:1 (cf. structures *26a* and *26b*). Optically active *27b* on the other hand reacts in a stereospecific manner. Only the Λ-*cis*-α-isomer is formed with (5R)-*27b*[81], containing the methyl group of the central chelate ring in an equatorial position. It is therefore not surprising that this ligand, fixing at the same time the topography as well as the absolute configuration of its complexes was proposed as a model for investigations of the stereoselective formation of mixed ligand complexes[83], but these investigations have not yet been accomplished.

2.2.3.2. Branched Ligands. Branched ligands are those with a basic structure *D* and *E*. For ligands of type *D* the nonterminal ligand atom becomes a center of chirality by the introduction of the ramification. The point of ramification is itself a center of chirality when the groups attached to this point are different in their nature. Geometrical and optical isomers are then possible. Different degrees of stereoselectivity can be expected in the formation of these isomers. In a similar way as with the corresponding tridentate ligands, *e.g.* aspartic acid, the three ligand atoms neighboring the branching point can only arrange in a facial manner. The absolute configuration of this arrangement is determined by the chirality of the ligand. Isomerism is still possible by the orientation of the remaining branch of the ligand. Fig. 2 gives the four possibilities which may be expected for such a ligand in its racemic form. These

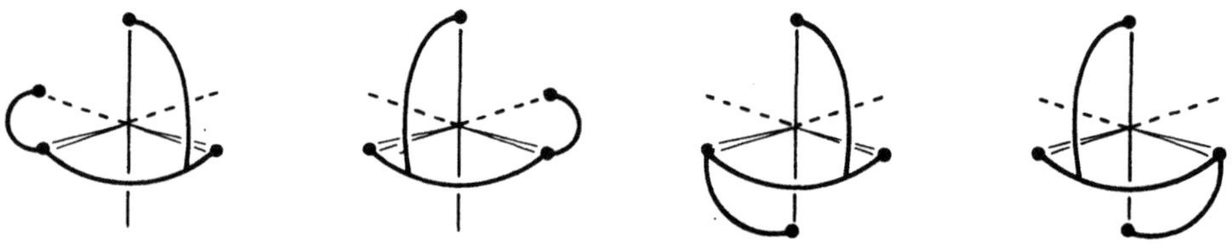

Fig. 2. Possibilities of isomeric complexes with tetradentate branched ligands (racemic)

possibilities will be doubled when one admits for each geometric form both left- and right-handed chirality of the asymmetric coordination atom regardless of the configuration of the branching point. In most cases this should lead to very high distortions in the geometry of the ligand, so that four theoretically possible diastereoisomeric forms should be very unstable.

The appearance of geometric and optical isomers with ligands of type *E* depends on some additional conditions. Both types of isomers may appear if
a) terminal coordination atoms are different in nature
b) the chelate rings are different in size
c) the chelate rings carry different substituents.

Diastereoselectivity is possible only in case c). Examples of the cases a) and b) are known and a definite selectivity concerning the possible geometrical isomers seems to exist as well for ligands with different terminal coordination atoms as for compounds with chelate rings of different size[84, 85]. Uehara and coworkers describe a certain number of optically active derivatives of nitrilotriacetic acid and its Co^{III}-complexes[86]. Koine and coworkers could obtain a restricted number of the possible isomers of mixed Co^{III}-amino acid/L- or D-alaminate-N,N-diacetic acid complexes[87]. The configurational chirality of these compounds has not been determined.

Information concerning the stereoselectivity with ligands of type *D* is also very limited. The stability measurements of mixed ligand complexes of N-(2-pyridyl)-L-aspartic acid and different amino acids with Cu^{2+} and Ni^{2+} show quite a high stereoselectivity, but they allow no conclusions as to the exact geometry of the tetradentate ligand in the complex[88]. Further studies with this type of ligands seem therefore necessary.

2.2.3.3. Macrocyclic Ligands. In a similar way as for linear tetramines, complexes with macrocyclic ligands of the type *30* and *31* offer certain interesting stereochemical aspects. A considerable number of isomers is possible by the presence of the four

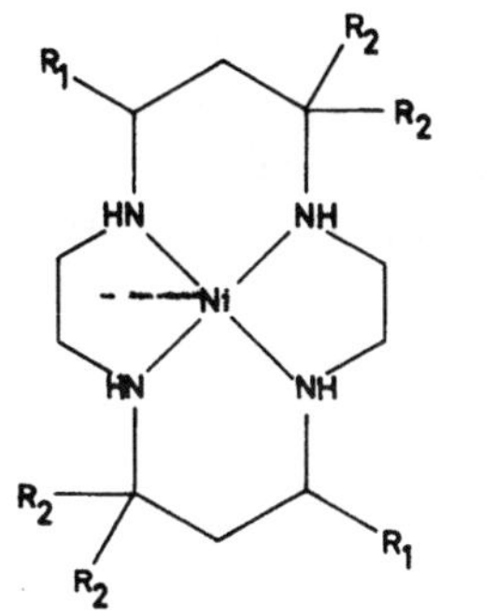

a) $R_1 = -CH_3; R_2 = -H$
b) $R_1 = R_2 = -CH_3$

30

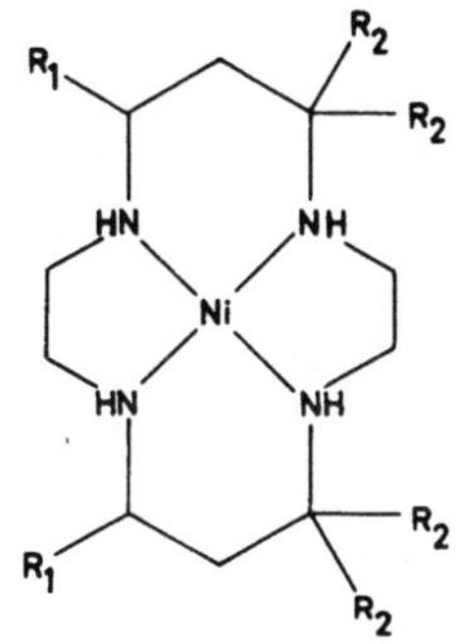

$R_1 = R_2 = -CH_3$

31

asymmetric nitrogen and the two asymmetric carbon atoms[89]. In a square planar arrangement, each of these complexes gives twenty diastereoisomers. Of these, sixteen exist as pairs of enantiomers and four show meso-configuration, giving a total of thirty-six isomers. The number of different topographies is, on the other hand, limited. Whereas open-chained tetramines can be obtained quite easily in the three mentioned topographies, the complexes of the macrocyclic ligands contain these latter most often in a square planar (*trans*) arrangement. A folded structure; possible only in one topography, can be obtained only for certain isomers by the reaction with a bidentate ligand.

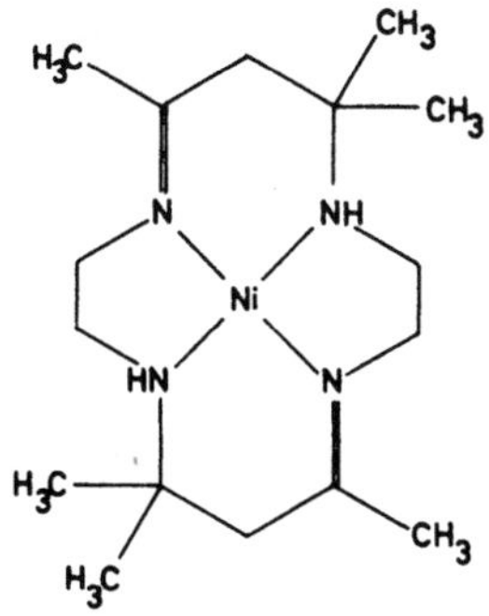

32

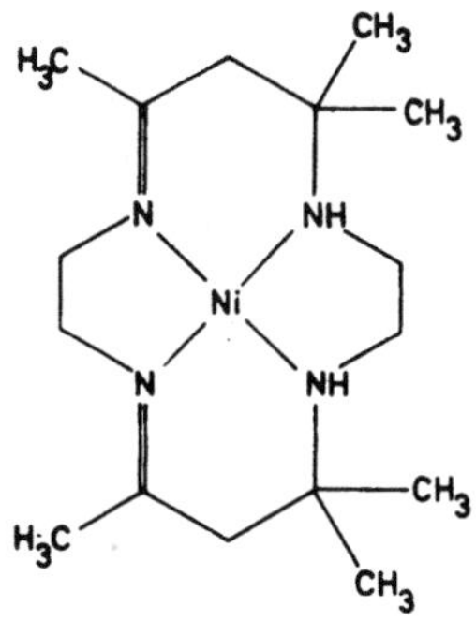

33

The very simple method for the synthesis of *32* and *33* by N. F. Curtis using $[Ni(en)_3](ClO_4)_2$ as starting material[90] made the compounds *30b* and *31* readily accessible by reduction. These compounds have therefore aroused great interest during the last years. By NMR-measurements it was shown that *32* exists as a meso- and a racemic form due to the presence of the two asymmetric nitrogen atoms[91]. The structure of the racemic compound and its configurational stability in acidic solution could be ascertained by partial separation of the optical antipodes[92]. Sledziewska studied the configurational stability of the asymmetric nitrogen atoms[93] and showed that the exchange of hydrogen atoms is 600 resp. 800 times faster in *32* resp. *33* than the loss of optical activity by inversion of the two nitrogen centers[94]. Wermeille and co-workers describe the stereoselective hydrogenation of optically active *32*[95]. The configuration of asymmetric carbon atoms is quantitatively determined by the configuration of the nitrogen atoms in the starting material, when the reaction is performed in acidic solution. By the application of this principle on meso-*30b* and rac-*31* it was possible to synthesize a certain number of new isomers[96]. It was further possible to make a proposal for absolute configuration of all optically active isomers so far known by comparing CD.-spectra with those of compounds with known absolute configuration[97, 98].

For the examples of diamagnetic square planar nickel(II)-complexes with macrocyclic tetramines, one observes that the position of the substituents is obviously of greater importance in determining the structure of the compound than the conformational effects of the chelate rings. This is surprising when one assumes that the interactions between substituents in axial position and ligands in the fifth or sixth coordination site are small in square planar complexes, as was found for platinum(II)-complexes (see p. 8). Nevertheless, the only example so far known with a methyl substituent in pseudoaxial position shows that such compounds may be obtained if several unfavorable conformational effects can be thus eliminated[98]. One may imagine such macrocyclic ligands being used as matrix compounds for stereoselective binding of secondary ligands, but such investigations have not yet been undertaken.

2.2.4. Penta- and Hexadentate Ligands

Except the propylenediaminetriacetic acid, recently described by Maricondi and Maricondi[99], no optically active ligands are known which react only as pentadentate ones. More frequent are ligands of the three structural types *F*, *G*, and *H*. According to the nature of the central metal atom and the reaction conditions these compounds may react as penta- or hexadentate ligands.

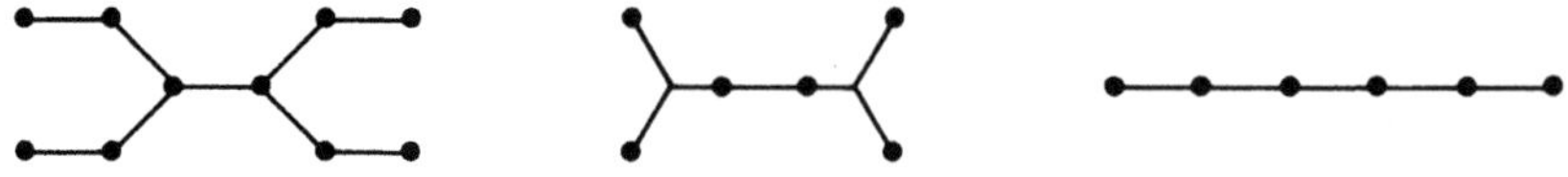

F *G* *H*

2.2.4.1. Linear Ligands. For only a very small number of ligands with structure H diastereoselective reactions are known. Theoretically four geometric arrangements α, α′, β and γ are possible (Fig. 3). None of these four possible geometries shows an

α α′ β γ

Fig. 3. Possible isomers of complexes with linear hexadentate ligands

improper rotation axis and all may therefore exist as enantiomers. By removing the terminal groups we may associate them to the corresponding isomers of complexes with linear tetradentate ligands: α and α′ to *cis*-α, β to *cis*-β and γ to *trans*. Considering the diastereoselectivity of these ligands we may expect a similar behavior for compounds substituted in the non-terminal chelate rings. This is confirmed by the few examples so far known. Dwyer and co-workers have prepared a series of complexes with ligands of the type *34*[100–102]. Whereas *34a* and *34b* give only a green complex

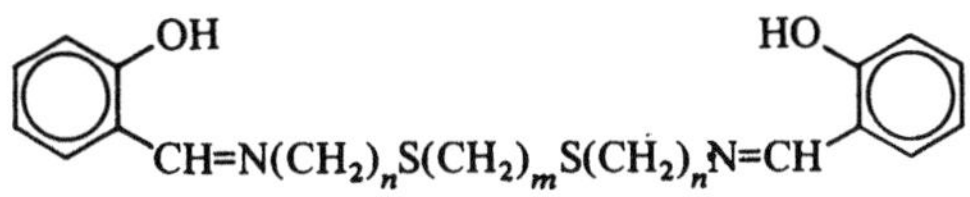

34

a) $n = \quad m = 2$ (EEE(sal)$_2$)
b) $n = 2\ m = 3$ (ETE(sal)$_2$)
c) $n = 3\ m = 2$

with cobalt(III), a green or a brown complex is obtained with *34c* depending on the reaction conditions. The α-configuration is generally attributed to the green complexes. For the brown compound a γ-structure seems more probable on the basis of CD-data data[103] than the α-structure first proposed[101]. 4-methyl-1,8-bis(salicylideneamino)-3,6-dithiaoctane [EPE(sal)$_2$] and 1,8-dimethyl-1,8-bis(salicylideneamino)-3,6-dithia-octane [PEP(sal)$_2$], two optically active ligands obtained from *27b* and *27c* give

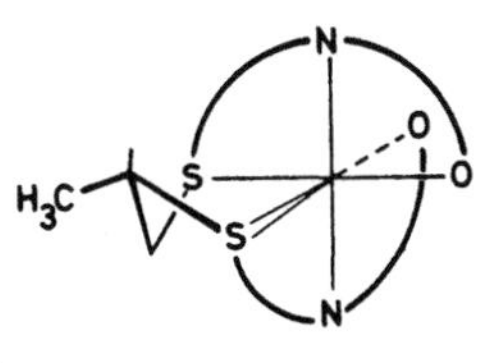

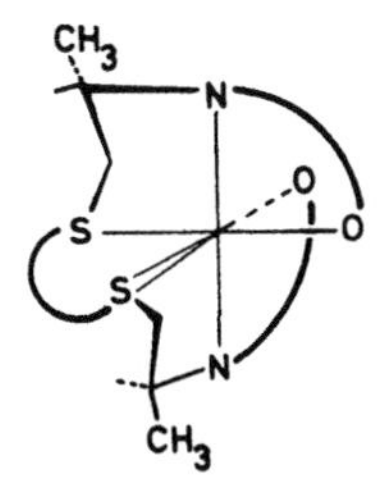

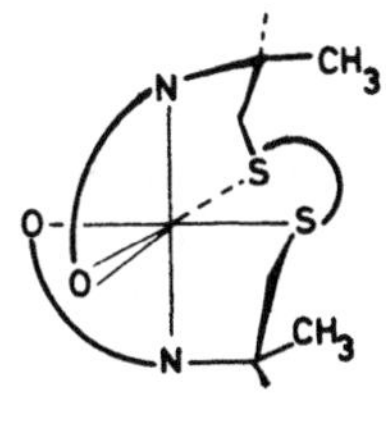

35 *36a* *36b*

complexes with α-topography and react in a diastereoselective way. For the ligand obtained from R(+)-EPE at least 85% of the more stable *35* is obtained, but it is likely that the selectivity is higher, the corresponding tetradentate ligand giving only one of the two diastereoisomers[81]. Complexes with (S,S)-PEP-(sal)$_2$ give an equilibrium mixture of the two diastereomeric forms *36a* and *36b*. 72% are found as the somewhat more stable exo-form *36a*[103]. The situation is therefore quite similar with respect to (S,S)-PEP. The Co^{III}-complex of this tetradentate ligand is a mixture containing 65% of the more stable exo-form. It seems therefore that the addition of two supplementary chelate rings is of little importance for the diastereoselectivity of these ligands. The geometric selectivity on the other hand seems more pronounced with hexadentate ligands.

2.2.4.2. Branched Ligands. In most cases of ligands of type *F* or *G*, the situation is quite simple. In general, ligands of these types react in a very selective way. The earliest known compounds were derivatives of propylenediamine and diaminocyclohexane. Dwyer and Garvan demonstrated the stereospecific coordination of 1,2-

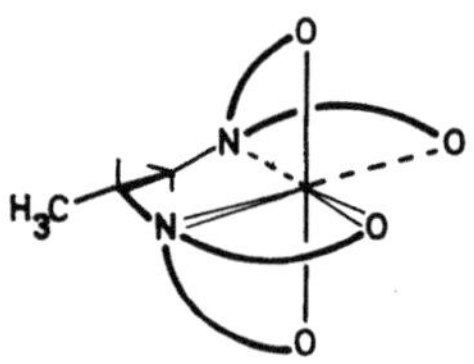

37a

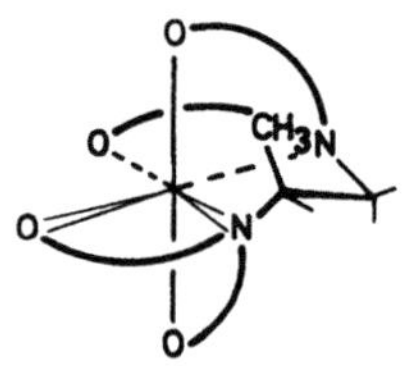

37b

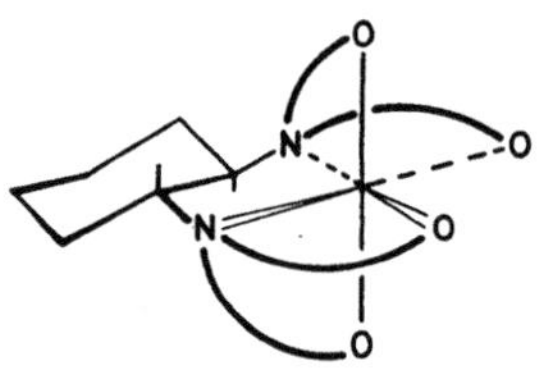

38

propylene diaminetetraacetic acid by forming the cobalt(III)-complexes of the racemic mixture of the ligand and by separating the optically active forms[104]. For the (R)-PDTA-complex these authors presume an energy difference of at least 3 kcal/mole between *37a* and *37b*[105]. This energy difference is due to the interaction between the axial methyl group and the chelate ring in the polar position. Blackmer and coworkers described the only case so far known of a PDTA-complex with an axial methyl group[106]. They showed that the completely reversible photomutarotation of $[Rh((-)\text{-}PDTA)]^-$ – a reaction previously observed by Dwyer and Garvan[105] – may only be explained by the transformation of *37a* in *37b*.

The configurational inversion of a complex with the corresponding derivative of diaminocyclohexane would lead to additional steric interactions. The diastereoisomeric complex of *38* with the enantiomeric configuration can therefore be excluded. The difference in configurational rigidity between PDTA- and CyDTA-complexes is further illustrated by the acetate group scrambling[107]. Whereas this reaction is not observed for $[Ni(CyDTA)]^{2-}$, $[Ni(EDTA)]^{2-}$ reacts with N-inversion as well as by Δ-Λ-isomerisation. With $[Ni(PDTA)]^{2-}$ the acetate group scrambling takes place only by N-inversion. In this reaction, the methyl group of the PDTA-ligand appears in a more or less pronounced axial position at the transition state.

The high stereoselectivity for the ligands mentioned above was observed with inert Co^{III}-complexes. This observation leads to the assumption that the corresponding labile complexes exist practically in one stereochemical form only[105, 108–110]. This stereoselectivity in labile complexes was used in different investigations, *e.g.*

the polarimetric titration of several cations[111], the stereoselective coordination of monodentate ligands in mixed-ligand complexes[112, 113], the polarimetric measurement of complex formation between diaminopolycarboxylates and alcaline ions[114, 115] the study of kinetic and mechanism of ligand exchange[116–118] and electron transfer reactions[119].

Certain ligands with the same framework as EDTA but with only one type of ligand atoms are also known[120, 121]. Gollogly and Hawkins have calculated an energy

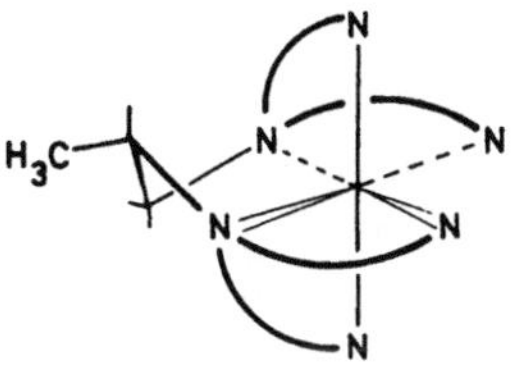

39a

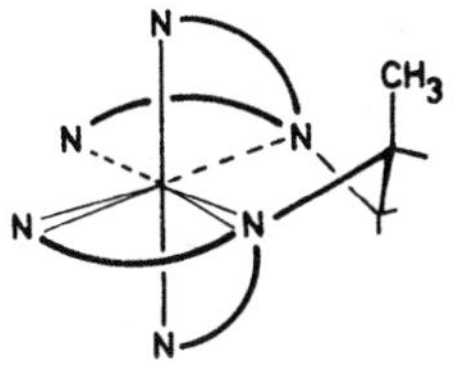

39b

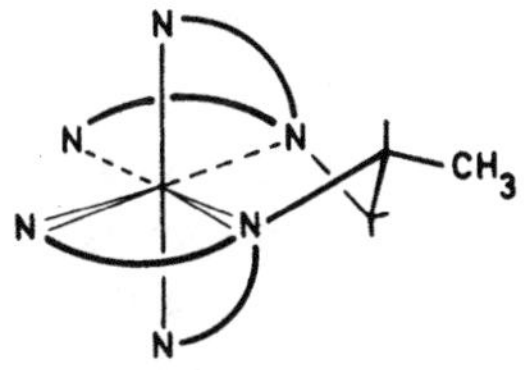

39c

difference of 8–10 kcal/mole between the two diastereoisomers *39a* and *39b* of N,N,N′,N′-tetrakis-(2-aminoethyl)-1,2-diaminopropane (mepenten)[122, 18]. This energy difference is very large. The optically active ligand gives therefore, like PDTA, complex compounds with only one generic configuration. These complexes have been obtained both by using the optically active ligand with known configuration[120] and by the separation of the mixture of the enantiomeric complexes *39a* and *39c* obtained by the reaction with the racemic ligand[121]. The absolute configuration has been proposed on the basis of chiroptical properties and confirmed by X-ray diffraction analysis.

Despite the results obtained by Emmenegger[123] showing that one chelate ring placed in the equatorial plane of the complex may be opened quite easily in a reaction with a monodentate ligand, it seems that the strain of this chelate ring is smaller in complexes of penten compared to EDTA. Contrary to the result with $[Ni(PDTA)]^{2-}$, it was not possible to coordinate 1-phenylethylamine in a stereoselective way with $[Ni((R)\text{-mepenten})]^{2+}$[113, 124].

As already mentioned for linear tetradentate ligands, a stereoselective coordination may be expected for branched hexadentate ligands which carry a substituent in

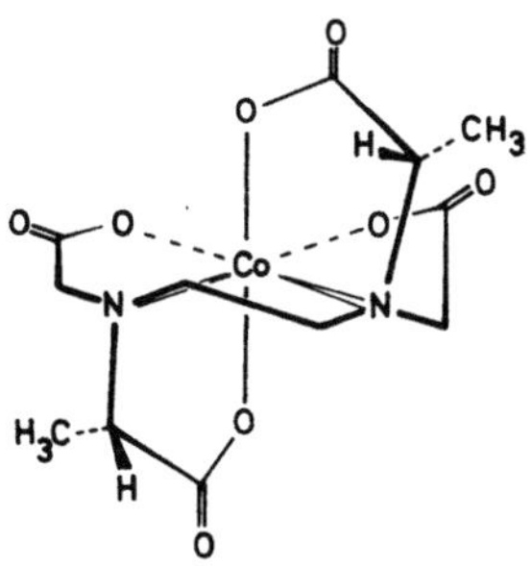

40a

Δ-exo-[Co((S,S)-EDDPA)]

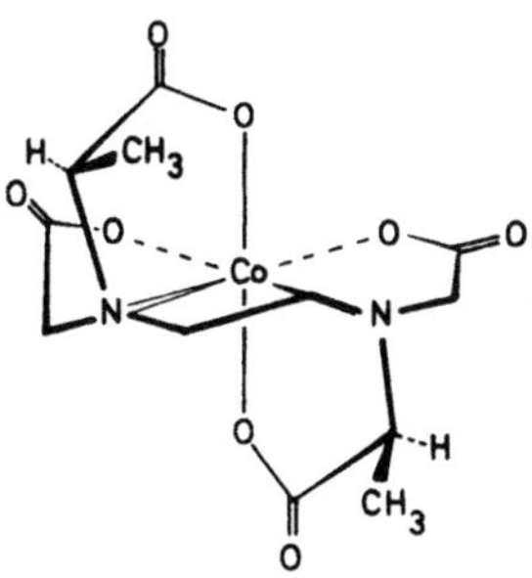

40b

Λ-endo[Co((S,S)-EDDPA)]

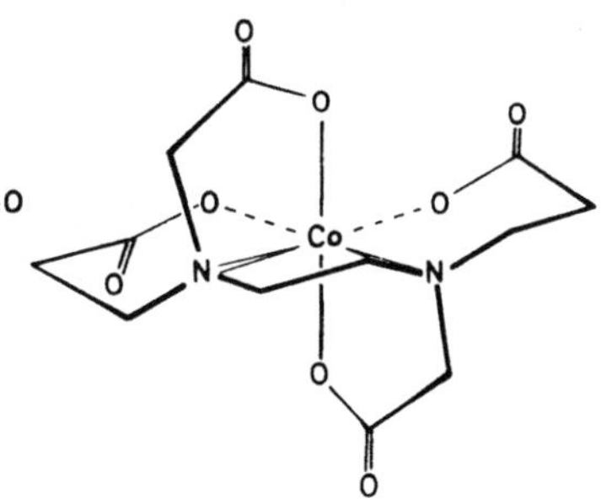

41

the central chelate ring. This selectivity is more or less pronounced when the substituents are placed in the external rings. In this way, geometrical selectivity is observed for the Co^{III}-complex of the ethylenediamine-N,N′-diacetic-N,N′-di-2-propionic acid; only the isomers with the substituted chelate rings in the polar positions are found. On the other hand, both possible diastereoisomers *40a* and *40b* of this geometrical form seem to exist at equilibrium[125]. An analogous selectivity between geometrical forms was observed for complexes with ethylenediamine-N,N′-diacetic-N,N′-dipropionic acid. In the stable form *41* the two six-membered rings are found in the equatorial plane[126]. Ligands with basic structure G giving chelate rings of different size behave in a similar way as *41* concerning the geometric selectivity. Furthermore these ligands react also in a diastereoselective manner due to the given chirality of the branching points. As seen for the corresponding tri- and tetradentate ligands, the configurational arrangement of the three ligand groups coming out from the branching point is determined by the configuration of the latter. Ethylenediamine-N,N-disuccinic acid (EDDS) contains two units of aspartic acid and was obtained from optically active aspartic acid with known configuration[127]. The absolute configura-

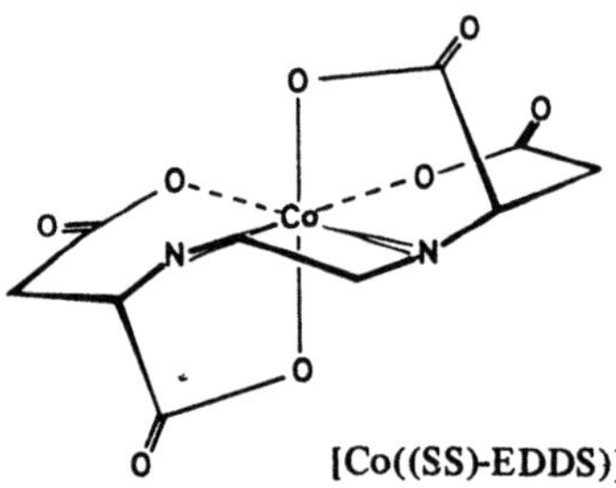

$[Co((SS)\text{-}EDDS)]^-$

42

tion of the complex is determined *42* if one assumes an equatorial position of the two six-membered chelate rings. The proposition of structure *42* was shown to be correct by X-ray diffraction analysis[128]. Complexes of such ligands present two asymmetric nitrogen atoms as supplementary elements of determined chirality. Their absolute configuration is fixed by the neighboring carbon atoms when the ligand acts as a hexadentate one. The asymmetric nitrogen atoms produce a considerable vicinal effect in the CD-spectra[127]. The higher tendency of EDDS to act as a hexadentate ligand which was also found for labile complexes is illustrated by crystal field[127] and NMR-data[129], and is attributed to the presence of the two six-membered chelate rings[130].

Another known ligand with the basic structure G is the ethylenediamine-N,N′-*bis*-[α-(2-hydroxy phenyl-acetic acid)] (H_4EDDHA) *43*. The interest of this ligand

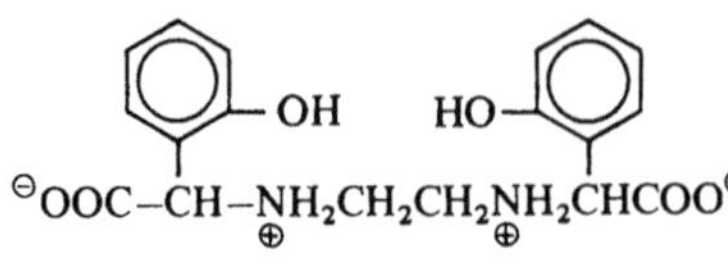

43

lies in the fact that it is so far the only known example existing as meso as well as racemic compound. Ryskiewich and Boka obtained a brown and a violet Fe^{III}-chelate by chromatographic separation[131]. The characterization of these two compounds becomes possible by separating the optical antipodes of the brown racemic substance[132]. From the antipodes the optically active ligands could be isolated. All attempts to separate the violet chelate were unsuccessful. This example shows that the separation of a racemic ligand reacting in a stereospecific way can be achieved not only by the separation of its inert complexes[81, 104, 133] but also of labile ones.

For the racemic ligand, the two geometric arrangements *44a* and *44b* are possible. In the complex with meso-ligand on the other hand, the two phenolic groups are

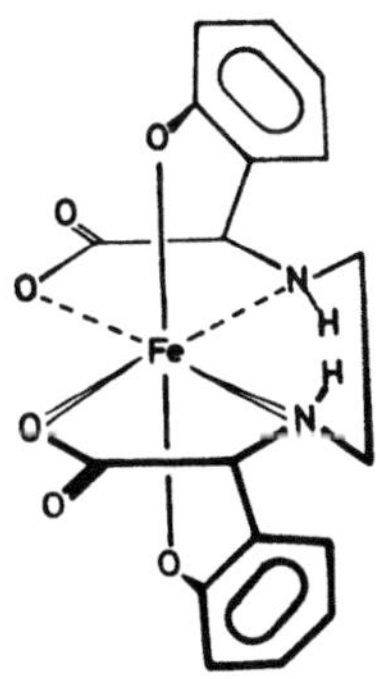

44a

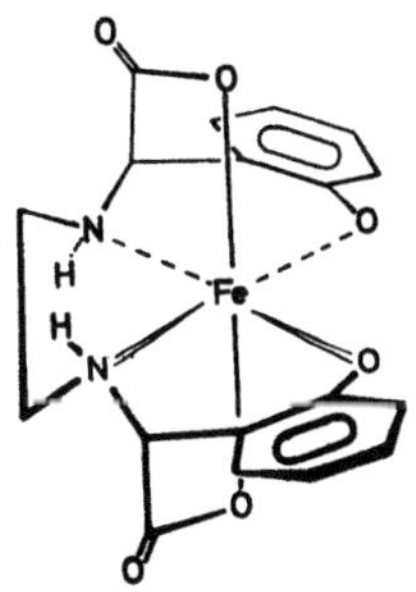

44b

$[Fe^{III}$-(R,R)-EDDHA$]^-$

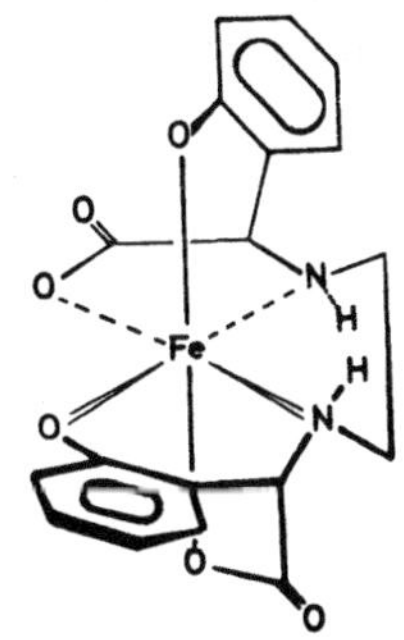

44c

$[Fe^{III}$-(R,S)-EDDHA)$]^-$

fixed in non-equivalent positions, only one geometric form *44c* therefore being possible. On the basis of the results obtained for EDDS-complexes, one may assume that the racemic ligand would prefer the arrangement *41b* showing both six-membered chelate rings in the equatorial position. A semi-quantitative indication as to the stability difference between the two forms *44a* and *44b* can be obtained by comparing the overall formation constants of the Fe^{III}-complexes with the meso- ($\log K_{ML}$ = 33,8) and the racemic ligand ($\log K_{ML}$ = 35,0)[132]. This difference corresponds to a ΔG-value of 1,65 kcal/mole. Taking this value as the energy necessary to place one six-membered chelate ring in a polar position (as is the case in the meso-complex) and considering that in *44a* both six-membered rings are found in this unfavorable position, we may conclude that the contribution of *44a* in the equilibrium $44a \rightleftharpoons 44b$ is

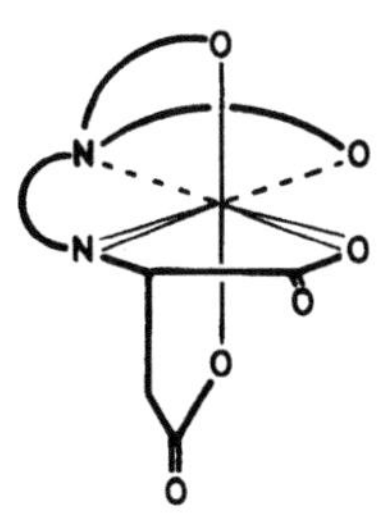

45a

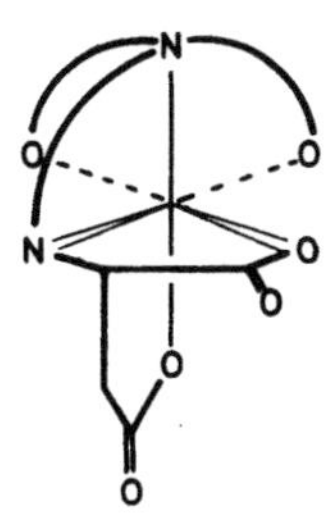

45b

less than 0,5 per cent. It may be interesting to note that the difference in the average crystal field stabilization between $[Co(EDTA)]^-$ and $[Co(EDDS)]^-$ is of the same magnitude[127].

An intermediate position between the EDTA and the EDDS-type ligands is occupied by (S)-ethylenediamine-N,N-diacetic-N′-monosuccinic acid (EDDAMS), a ligand described by Legg and Neal[134]. The ligand reacts in a diastereoselective way due to the presence of the aspartic acid unit, but allows the formation of two geometric isomers *45a* and *45b*. The difference between these two isomers lies in the position of the six-membered ring. Only *45b*, with the six-membered ring in the equatorial plane is obtained. This result confirms that considerable strain must exist in an equatorial arrangement of three five-membered rings and that this strain can be eliminated by introduction of one six-membered chelate ring.

3. Conclusions and Outlook

Studies of the stereochemical features in the reactions of metal ions with optically active ligands have aroused great interest during the last years but are still far from being complete. A systematic discussion is not yet possible. Despite this, the known examples furnish sufficient indications as to the stereochemical requirements of ligands with a definite geometry in the coordination entity. The parameters which have to be considered in this context are:

a) the nature of the ligand atoms
b) the number and size of chelate rings
c) the way the chelate rings are connected
d) the type and position of substituents
e) the nature of the metal atom.

The parameters a), b) and c) determine the topography of the complex. One may consider a complex formed by a multidentate ligand as a combination of several sub-units comprizing three donor atoms and two condensed chelate rings. A complex with a hexadentate ligand of type EDTA, for example, gives six possible combinations of such sub-units. Such a tridentate unity offers two structural possibilities as shown in Fig. 4. Different chemical characteristics of the currently employed ligands are

peripheric facial

Fig. 4. Possible arrangement of a linear tridentate sub-unit of a multidentate ligand

known to favor either a peripheric or a facial arrangement. The following may be mentioned.

1. Peripheric:
   - oxygen as central ligand atom
   - six-membered chelate rings
   - a central nitrogen atom with sp^2-hybridization
   - peptide groups
2. Facial:
   - branched ligands
   - two five-membered chelate rings with terminal carboxylate groups
   - sulfur or phosphorous as central ligand atoms.
3. Borderline case:
   - ligands of diethylenetriamine-type.

These characteristics take into account only steric factors of the basic framework of a ligand and may therefore give only a first approach to the geometry of the most stable form of an octahedral complex.

The parameter d) – nature and position of substituents of a ligand – may sometimes modify the basic geometry or the topography of the complex, but it is the determining factor for a definite chirality. It may determine the configuration of the complex, the conformation of chelate rings as well as the absolute configuration of asymmetric ligand atoms.

The influence of the nature of the metal ion on the stereochemistry of its environment in octahedral complexes is not well known. Most of the studies have been performed with compounds of cobalt(III). It would be useful to get more information about the influence of crystal field parameters, charge and ionic radii on the stereochemistry of metal complexes.

The limited number of optically active ligands has certainly been a real obstacle for a systematic study of diastereoselectivity in metal complexes. Usually these ligands are synthesized, starting with readily accessible natural substances or with compounds which can easily be separated into antipodes. The synthesis is then performed in such a way that the absolute configuration of the asymmetric center remains unchanged.

One would expect that the isolation of optically active ligands by separation of their corresponding complexes would find wider application in the future. Ligand exchange chromatography, which applies stereoselective formation of binary complexes on the absorbent, may be a good example of such a procedure. In this way one may hope that a certain number of ligands will soon be available, able to form complexes with predictable and unique topography and chirality.

Acknowledgement. The author would like to thank C. Bernhard and P. Baccini for their suggestions and help in preparing the manuscript.

4. References

1) Corey, E. J., Bailar, J. C., Jr.: J. Amer. chem. Soc. *81*, 2620 (1959).
2) Sargeson, A. M.: in Transition Metal Chemistry (R. L. Carlin, ed.), Vol. 3, pp. 303–343. New York: Marcel Dekker, Inc. 1966.
3) Sargeson, A. M.: in Chelating Agents and Metal Chelates (F. P. Dwyer and D. P. Mellor, eds.), chap. 5. New York: Academic Press 1964.
4) Dunlop, J. H., Gillard, R. D.: in Adv. in Inorg. and Radiochem. (H. J. Emeleus and A. G. Sharpe, eds.), Vol. 9, pp. 185–215. New York: Academic Press 1966.
5) Gillard, R. D.: Inorg. chim. Acta, Rev. *1*, 69 (1967).
6) Bernaüer, K.: in Metal Ions in Biological Systems (H. Sigel, ed.) Vol. 3, pp. 117–128. New York: Marcel Dekker, Inc. 1974.
7) Brubaker, G. R.: Chem. Educ. *51*, 608 (1974).
8) Breslow, R., Overman, L. E.: J. Amer. Chem. Soc. *92*, 1075 (1970).
9) Österberg, R.: Coord. Chem. Rev. *12*, 309 (1974).
10) Coleman, J. E.: in Catalysis, Progress in Research (F. Basolo and R. L. Burwell, Jr., eds.), pp. 3–41. New York: Plenum Press 1973.
11) Ogston, A. G.: Nature *162*, 963 (1948).
12) Fischer, E.: Ber. *27*, 2985 (1894).
13) Ritsma, J. H.: Thèse. University Groningen 1973.
14) Froebe, L. R., Yamada, S., Hidaka, J., Douglas, B. E.: J. Coord. Chem. *1*, 183 (1971). Yamada, S., Hidaka, J., Douglas, B. E.: Inorg. Chem. *10*, 2187 (1971).
15) Brubaker, G., Massura, J. G.: J. Coord. Chem. *3*, 251 (1974).
16) Rosenblatt, F., Schleede, A.: Liebigs Ann. *505*, 51 (1933).
17) Theilacker, W.: Z. Anorg. Chem. *234*, 161 (1937).
18) Hawkins, C. J.: Absolute Configuration of Metal Complexes. New York: Wiley Interscience 1971.
19) Basolo, F., Chen, Y. R., Murmann, R. K.: J. Amer. chem. Soc. *76*, 956 (1954).
20) Majer, J., Dvorakova, E.: Chem. Zvesti *17*, 402 (1963).
21) Majer, J., Novak, V., Svicekova, M.: Chem. Zvesti *18*, 481 (1964).
22) Saburi, M., Yoshikawa, S.: Bull. chem. Soc. Jap. *45*, 1443 (1972).
23) Goto, M., Matsushita, H., Saburi, M. Yoshikawa, S.: Inorg. Chem. *12*, 1498 (1973).
24) Güntert, B., Claude, S., Bernauer, K.: Helv. *58*, 780 (1975).
25) Dwyer, F. P., Garvan, F. L., Shulman, A.: J. Amer. chem. Soc. *81*, 290 (1959).
26) Dwyer, F. P., MacDermott, T. E., Sargeson, A. M.: J. Amer. chem. Soc. *85*, 2913 (1963).
27) Erickson, L. E., McDonald, J. W.: Howie, J. K., Clow, R. P.: J. Amer. chem. Soc. *90*, 6371 (1968).
28) Ho, F. F.-L., Erickson, L. E., Watkins, S. R., Reilley, C. N.: Inorg. Chem. *9*, 1139 (1970).
29) Ho, F. F.-L., Reilley, C. N.: Anal. Chem. *41*, 1835 (1969).
30) Gollogly, J. R., Hawkins, C. J.: Inorg. Chem. *11*, 156 (1972).
31) Saito, Y., Noumra, F., Marumo, F. M.: Bull. chem. Soc. Jap. *41*, 530 (1968); ibid *42*, 1016 (1969).
32) Pajunen, A.: Suomen Kenistilehti B *41*, 232 (1968); ibid B *42*, 15 (1969).
33) Appleton, T. G., Hall, J. R.: Inorg. Chem. *9*, 1807 (1970); ibid *10*, 1717 (1971).
34) Sarneski, J. E., Reilley, C. N.: Inorg. Chem. *13*, 977 (1974).
35) Freeman, H. C.: Adv. Protein Chemistry *22*, 257 (1967).
36) Weeks, C. M., Cooper, A., Norton, D. A.: Acta Cryst. B *25*, 443 (1969).
37) Gillard, R. D., Mason, R., Payne, N. C., Robertson, G. B.: J. chem. Soc. A, 1864 (1969).
38) Wellmann, K. M., Mungall, W., Mecca, T. G., Hare, C. R.: J. Amer. chem. Soc. *89*, 3647 (1967).
39) Blount, J. F., Freeman, H. C., Sargeson, A. M., Turnbull, K. R.: Chem. Commun. 324 (1967).
40) Zompa, L. J.: Chem. Commun. 783 (1969).
41) Freeman, W. A., Liu, C. F.: Inorg. Chem. *7*, 764 (1968).

42) Liu, Ch. F., Ibers, J. A.: Inorg. Chem. *8*, 1911 (1962).
43) Freeman, W. A., Liu, Ch. F.: Inorg. Chem. *9*, 1191 (1970).
44) Legg, J. I., Cooke, D. W. J. Amer. chem. Soc. *89*, 6854 (1967).
45) Gillard, R. D., McKenzie, E. D., Mason, R., Robertson, G. B.: Nature *209*, 1347 (1966).
46) Gillard, R. D., Harrison, P. M., McKenzie, E. D.: J. chem. Soc. A. 618 (1967).
47) Gillard, R. D.: Inorg. Chim. Acta Rev. 69 (1967).
48) Gillard, R. D., McKenzie, E. D., Mason, R., Robertson, G. B.: Coord. Chem. Rev. *1*, 263 (1966).
49) Gillard, R. D., Spencer, A.: J. chem. Soc. Dalton 902 (1972).
50) Michailidis, M. S., Martin, R. B.: J. Amer. chem. Soc. *91*, 4683 (1969).
51) Browing, J. G., Gillard, R. D., Lyons, J. R., Mitchell, P. R., Phipps, D. A.: J. chem. Soc. Dalton 1815 (1972).
52) Brubaker, G. R., Schaefer, D. P., Worrell, J. H., Legg, J. I.: Coord. Chem. Rev. *7*, 161 (1971).
53) Buckingham, D. A., Marzilli, P. A., Sargeson, A. M.: Inorg. Chem. *6*, 1032 (1967).
54) Sargeson, A. M., Searle, G. H.: Inorg. Chem. *6*, 787 (1967).
55) Goto, M., Saburi, M., Yoshikawa, S.: Inorg. Chem. *8*, 358 (1969).
56) Asperger, R. G., Liu, Ch. F.: Inorg. Chem. *4*, 1395 (1965).
57) Asperger, R. G., Liu, Ch. F.: Inorg. Chem. *6*, 796 (1967).
58) Brubaker, G. R., Euler, R. A.: Inorg. Chem. *11*, 2357 (1972).
59) Saburi, M., Sawai, T., Yoshikawa, S.: Bull. chem. Soc. Jap. *45*, 1086 (1972).
60) Goto, M., Okubo, A., Sawai, T., Yoshikawa, S.: Inorg. Chem. *9*, 1488 (1970).
61) Bosnich, B., Kneen, W. R.: Inorg. Chem. *9*, 2191 (1970).
62) Cragel, J., Jr., Brubaker, J. R.: Inorg. Chem. *11*, 303 (1972).
63) Häfeli, J.: Thesis 1975. Neuchâtel.
64) Asperger, R. G.: Inorg. Chem. *8*, 2127 (1969).
65) Inorg. Chem. *9*, 1 (1970).
66) Cahn, R. S., Ingold, C. K., Prelog, V.: Angew. Chem. Int. Ed. Engl. *5*, 385 (1966).
67) Harrington, P. C., Linke, S., Alexander, M. D.: Inorg. Chem. *12*, 168 (1973).
68) Stöckli-Evans, H.: Helv., to publish.
69) Warner, L. G., Busch, D. H.: Coord. Chem. – Papers presented in Honor of Professor John C. Bailar, Jr., New York: Plenum Press 1969.
70) DeHayes, L. J., Parris, M., Busch, D. H.: Chem. Commun. 1398 (1971).
71) DeHayes, L. J., Busch, D. H.: Inorg. Chem. *12*, 2010 (1973).
72) Legg, J. I., Cooke, D. W.: Inorg. Chem. *4*, 1576 (1965).
73) Garnett, P. J., Watts, D. W., Legg, J. I.: Inorg. Chem. *8*, 2534 (1969).
74) Liu, C. F.: Inorg. Chem. *3*, 680 (1964).
75) Garnett, P. J., Watts, D. W.: Inorg. Chim. Acta *8*, 293 (1974).
76) Sudmeier, J. L., Occupati, G.: Inorg. Chem. 7, 2524 (1968).
77) Schoenberg, L. N., Cooke, D. W., Liu, C. F.: Inorg. Chem. *7*, 2386 (1968).
78) Bosnich, B., Kneen, W. R., Phillip, A. T.: Inorg. Chem. *8*, 2567 (1969).
79) Worrell, J. H., Busch, D. H.: Inorg. Chem. *8*, 1563 (1969).
80) Worrell, J. H., Busch, D. H.: Inorg. Chem. *8*, 1572 (1969).
81) Worrell, J. H., McDermott, T. E., Busch, D. H.: J. Amer. chem. Soc. *92*, 3317 (1970).
82) Bosnich, B., Phillip, A. T.: J. Chem. Soc. (A) 264 (1970).
83) Worrell, J. H., McDermott, T. E., Busch, D. H.: Chem. Commun. 661 (1969).
84) Chu, M. W. S., Cooke, D. W., Liu, Ch. F.: Inorg. Chem. 7, 2453 (1968).
85) Schneider, P. W., Collman, J. P.: Inorg. Chem. 7, 2010 (1968).
86) Uehara, A., Kyuno, E., Tsuchiya, R.: Bull. chem. Soc. Jap. *44*, 1552 (1971).
87) Koine, N., Sakota, N., Hidaka, J., Shimura, Y.: Bull. chem. Soc. Jap. *43*, 1737 (1970).
88) Nakon, R., Rechani, P. R., Angelici, R. J.: Inorg. Chem. *12*, 2431 (1973).
89) See Ref.[69] p. 3ff.
90) Curtis, N. F., House, D. A.: Chem. and Ind. 1708 (1961).
91) Warner, L. G., Rose, N. J., Busch, D. H.: J. Amer. chem. Soc. *90*, 6938 (1968).
92) Warner, L. G., Rose, N. J., Busch, D. H.: J. Amer. chem. Soc. *89*, 703 (1967).
93) Sledziewska, E.: Bull. Acad. Polon. Sci., Ser. sc. chim. *20*, 49 (1972).

94) Sledziewska, E., Vonderschmitt, D.: Bull. Acad. Polon. Sci. Ser. sc. chim. *20*, 117 (1972).
95) Wermeille, M., Sledziewska, E., Bernauer, K.: Helv. *57*, 180 (1974).
96) Wermeille, M.: Thesis. Neuchâtel 1974.
97) Curtis, N. F., Swann, D. A., Waters, T. N., Maxwell, J. E.: J. Amer. chem. Soc. *91*, 4588 (1969).
98) Curtis, N. F., Swann, D. A., Waters, T. N.: J. Chem. Soc. Dalton 1963 (1973).
99) Maricondi, C. W., Maricondi, Ch.: Inorg. Chem. *13*, 1110 (1974).
100) Dwyer, F. P., Lions, F.: J. Amer. chem. Soc. *72*, 1545 (1950).
101) Dwyer, F. P., Gill, N. S., Gyarfas, E. C., Lions, F.: J. Amer. chem. Soc. *74*, 4188 (1952).
102) Dwyer, F. P., MacDermott, T. E.: J. Amer. chem. Soc. *85*, 2916 (1963).
103) Bosnich, B., Phillip, A. T.: J. Amer. chem. Soc. *90*, 6352 (1968).
104) Dwyer, F. P., Garvan, F. L.: J. Amer. chem. Soc. *81*, 2955 (1959).
105) Dwyer, F. P., Garvan, F. L.: J. Amer. chem. Soc. *83*, 2610 (1961).
106) Blackmer, G. L., Sudmeier, J. L., Thibedeau, R. N., Wing, R. M.: Inorg. Chem. *11*, 189 (1972).
107) Erickson, L. E., Young, D. C., Ho, F. F.-L., Watkins, S. R., Terrill, J. B., Reilley, C. N.: Inorg. Chem. *10*, 441 (1971).
108) Day, R. J., Reilley, Ch. N.: Anal. Chem. *37*, 1326 (1965).
109) Callahan, K. P., Ostrom, G. A., Wing, R. M.: Inorg. Chem. *9*, 2605 (1970).
110) Wing, R. M. Callahan, K. P.: Inorg. Chem. *8*, 2303 (1969).
111) Palma, R. J., Reinbold, P. E., Pearson, K. H.: Chem. Commun. 254 (1969).
112) Humbel, F., Vonderschmitt, D., Bernauer, K.: Helv. *53*, 1983 (1970).
113) Vonderschmitt, D., Jeanneret, M. F., Bernauer, K.: Helv. *54*, 297 (1971).
114) Carr, J. D., Schwartzfager, D. G.: Anal. Chem. *43*, 583 (1971).
115) Carr, J. D., Schwartzfager, D. G.: Anal. Chem. *43*, 1520 (1971).
116) Reinbold, P. E., Pearson, K. H.: Inorg. Chem. *9*, 2325 (1970).
117) Bosnich, B., Dwyer, F. P., Sargeson, A. M.: Nature *186*, 966 (1960).
118) Carr, J. D., Torrance, K., Cruz, C. J., Reilley, C. N.: Anal. Chem. *39*, 1358 (1967).
119) Jm, Y. A., Busch, D. H.: J. Amer. chem. Soc. *83*, 3362 (1961).
120) Gollogly, J. R., Hawkins, C. J.: Chem. Commun. 873 (1966).
121) Yoshikawa, Y., Fuji, E., Yamasaki, E.: Bull. chem. Soc. Jap. *45*, 3451 (1972).
122) Gollogly, J. R., Hawkins, C. J.: Austr. J. Chem. *20*, 2395 (1967).
123) Emmenegger, F. P., Schwarzenbach, G.: Helv. *49*, 625 (1966).
124) Vonderschmitt, D.: private communication.
125) Williams, D. H., Angus, J. R., Steele, J.: Inorg. Chem. *8*, 1374 (1969).
126) Byers, W., Douglas, B. E.: Inorg. Chem. *11*, 1470 (1972).
127) Neal, J. A., Rose, N. J.: Inorg. Chem. *7*, 2405 (1968).
128) Woodward, L. M.: Thesis. University of Washington 1970.
129) Grant, M. W. Dodgen, H. W., Hunt, J. P.: J. Amer. chem. Soc. *93*, 6828 (1971).
130) Neal, J. A., Rose, N. J.: Inorg. Chem. *12*, 1226 (1973).
131) Ryskiewich, D. P., Boka, D.: Nature 473 (1962).
132) Bernauer, K., Colomb, G.: to publish.
133) Bosnich, B., Wild, S. B.: J. Amer. chem. Soc. *92*, 459 (1970).
134) Legg, J. I., Neal, J. A.: Inorg. Chem. *12*, 1805 (1973).

Received March 13, 1975

Mechanistic Aspects of the Photochemical Reactions of Coordination Compounds

Prof. Mark S. Wrighton[1)]

Department of Chemistry, Massachusetts Institute of Technology, Cambridge, Massachusetts 02139, U.S.A.

Contents

Introduction

Increasing interest[2] in the photochemistry of transition metal containing molecules prompts this survey of the important developments in understanding the chemical transformations resulting from electronic excitation of such molecules. Intense and continuous efforts[3] have been applied to elucidation of the electronic structure of transition metal complexes, and it thus seems appropriate to subject the emerging theories to a test of their applicability to the chemical properties of the excited states accessible by absorption of light of optical energies. More importantly, attention is drawn to this area of chemistry by the nature of the chemical reactions possible and their potential application in energy conversion, synthesis, catalysis, photochromism, photography and other imaging systems, and optical devices in general. The specific aim of this article is to acquaint the reader with the results of the last few years which tend to lay the foundation for new principles or those results which are sufficiently interesting and misunderstood that they deserve further effort. No attempt will be made to discuss separately the electronic structures of coordination compounds except in very general and simple terms. Nor is this article any attempt to review every photoreaction of coordination compounds. Finally, excited state processes such as energy transfer, luminescence, etc., will not be discussed in separate sections but will be referred to as they pertain to reactions.

The photochemistry of transition metal complexes of various d^n electronic configurations will be discussed. Six-coordinate complexes of the d^3 and low-spin d^6 configuration and four-coordinate low-spin d^8 complexes have received most attention as these are the configurations which are most thermally substitution inert[4] and undergo a large degree of labilization upon population of their lowest ligand field (LF) excited states.[5] Highly covalent organometallic complexes, including metal carbonyls, provide numerous examples of photoreactions where the metal can be viewed as having d^4, d^5, d^6, d^7, d^8, and d^9 electronic configurations. In addition to LF excited states several different types of charge-transfer (CT) excited states and intraligand (IL) excited states are accessible at optical energies for transition metal compounds. It will be demonstrated that reasonable sense can be drawn from a correlation of the available excited states and the observed chemistry, but the overall organization of this paper rests in the reaction types: substitution, redox, intraligand, and metal-metal bond cleavage.

I. Photosubstitution Chemistry

Ligand substitution remains the most well-studied of the photoreactions of coordination compounds. Virtually every thermally inert (and some labile) type of complex has received some study. These include complexes of 4, 5, 6, 7, and 8 coordination having both unusual and regular geometries and among these we find examples of nearly every d^n electronic configuration. Prior to outlining the questions of mechanistic significance some of the possible uses of ligand photosubstitution will be briefly mentioned. First, ligand substitution can be used to synthesize new metal complexes. To do this photochemically may yield new complexes inherently inaccessible by thermal substitution because of the stereochemistry, lack of any thermal activity, or because of competing thermal reactions involving other portions of the complex. Specific examples of these will be discussed in the sections below. Photosubstitution chemistry has been shown to be an important aspect of photoinitiated transition metal catalysis.[6] Indeed, all three of the general steps of catalytic cycle, (1)–(3), can potentially be photoaccelerated and (1) and (3) are both types of photosubstitution. In these and other cases the mechanism for simple ligand substitution may

$$\text{Substrate} + \text{Metal Complex} \xrightarrow{\Delta \text{ or } h\nu} [\text{Substrate} - \text{Metal}] \tag{1}$$

$$[\text{Substrate} - \text{Metal}] \xrightarrow{\Delta \text{ or } h\nu} [\text{Transformed Substrate} - \text{Metal}] \tag{2}$$

$$[\text{Transformed Substrate} - \text{Metal}] \xrightarrow{\Delta \text{ or } h\nu} \text{Transformed Substrate} + \text{Metal Complex} \tag{3}$$

involve a truly coordinatively unsaturated intermediate capable of being intercepted by nucleophiles which leads to simple substitution or the intermediate may be susceptible to oxidative addition by a substrate also present in the medium. In cases where coordinative unsaturation can be generated the nature and stereochemical position of the ligand dissociated becomes the point of concern and this information significantly adds to our understanding of the structure and chemistry of the incipient coordinatively unsaturated intermediate. Such intermediates are omnipresent in homogenous catalysis. Since photosubstitution reactions may involve large changes in optical spectra such chemistry represents a large reservoir of potential photochromic systems. Large spectral changes can be expected to result from ligand substitutions which substantially change the average LF strength or from the loss or incorporation of a ligand involved in CT absorptions. The release of particular ligands may have other physical consequences which may have use in light sensitive devices. Such physical changes include variation in pH and in the number of charged species. Finally, in fundamental terms, photosubstitution reactions serve, to some extent, as a chemical test for our theories concerning the metal-ligand bonding and the nature of the excited states of the complex.

Most intense activity mechanistically has focussed on the correlation of the excited state achieved and the chemistry observed. More precisely, the question is what is the reactive excited state? Secondly, how does ligand substitution occur subsequent to electronic excitation? This second question has been asked much less often than the first, but each is of crucial importance in understanding the chemistry.

Answering the question concerning the reactive state depends on knowing the excited states which are available and such information comes together from the theories of electronic structure and experimental measurements such as absorption spectra and electron spectroscopy. Even though this work is extensive and probably definitive, the question of which state(s) lead to reaction is very difficult to answer and even now only a few systems are well studied enough that no reasonable doubt remains as to the identification of the reactive state(s). Some attempts to make predictions concerning the course of photosubstitution reactives have been made. The various predictive theories will be put into experimental perspective below, but some historical development may be useful here. The first predictions involved Cr(III) complexes where empirical observations led to "Adamson's rules" which enabled prediction of the leaving group in a series of substituted Cr(III)-amine complexes.[7] Several years later, Zink[8] elaborately rationalized the "rules" in terms of molecular orbital (MO) and LF theory. In certain key cases Zink's model and the "rules" are at variance and it appears that the MO approach can place the correct predictions of the "rules" on a firm foundation and can lend useful and practical comments where the "rules" fail. At about the same time as Zink's work appeared, Wrighton, Gray, and Hammond[5] published a paper outlining a model for the substitutional reactivity of LF excited states in d^3, low-spin d^6, and low-spin d^8 complexes. This model is, by comparison to Zink's, qualitative but still based on LF and MO arguments. Zink has subsequently extended[9] his model to certain lowspin d^6 systems and has fully detailed the Cr(III) story. Even more recently, Rösch, Messmer, and Johnson[10] have pointed out that their calculations using the self-consistent field-X alpha (SCF-Xα) scattered wave method will account for certain features of the photosubstitution behavior of coordination compounds.

Return now to the questions surrounding the actual sequence of events leading to substitution following population of the reactive state. As in thermal substitution mechanisms it is appropriate to determine whether a dissociative or an associative mechanism obtains. Certainly, this point is the one most often clarified, but other aspects also deserve some scrutiny. These include the possibility of acid-base equilibria in the excited state, isomerization of potentially ambidentate ligands, the extent to which the extruded ligand is electronically or vibrationally excited, the degree of molecular distortion upon population of the reactive state and the possibility of competing chemical processes which may be influenced by the environment or by structural modifications of the molecule.

A. Photosubstitution of Cr(III) Complexes

Complexes of Cr(III) have received more detailed study than those of any other metal. Generally, these d^3 complexes are rather substitutionally inert at 25 °C but do undergo very efficient photosubstitution upon population of the low lying LF excited states. It is interesting to point out now that specific results from Cr(III) photochemistry are not likely to be very useful generally as few other d^3 mononuclear complexes are thermally stable enough to be of interest to photochemists except perhaps mechanistically. Nonetheless, Cr(III) complexes will serve nicely to introduce the important theories and some interesting experimental results and techniques applicable to other cases. We begin first with a discussion of the nature of the reactive state(s) and then a discussion of the mechanism of ligand substitution from this state.

1. The Substitutionally Reactive State of Cr(III) Complexes

The electronic structural theory for simple, six-coordinate Cr(III) complexes is well founded.[11] For the O_h case the ground state electronic configuration is t_{2g}^3 giving rise to the $^4A_{2g}$ ground state. The inertness of the ground state may be associated with this half-filled t_{2g} level. Excited states at low energy are of two fundamentally different types. The first type arises from spin-pairing as indicated in Scheme 1 and does not involve changes in 10Dq. These one-electron transitions correspond to weak, spin-forbidden absorptions found in all Cr(III) complexes near 13 kK since they are

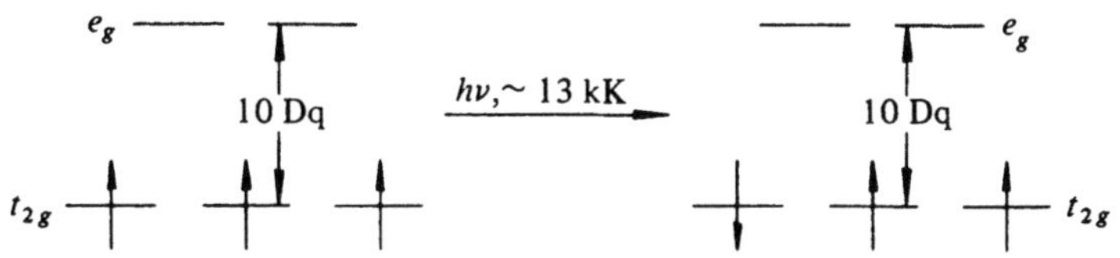

Scheme 1

essentially independent of LF strength. The ^2E_g, $^2T_{1g}$, and $^2T_{2g}$ states are the excited states arising from the t_{2g}^3 configuration. The second type of LF excited states are those derived from the $t_{2g}^2e_g^1$ one-electron configuration. The one-electron transition in this case, Scheme 2, does involve 10Dq and hence the energetic position of

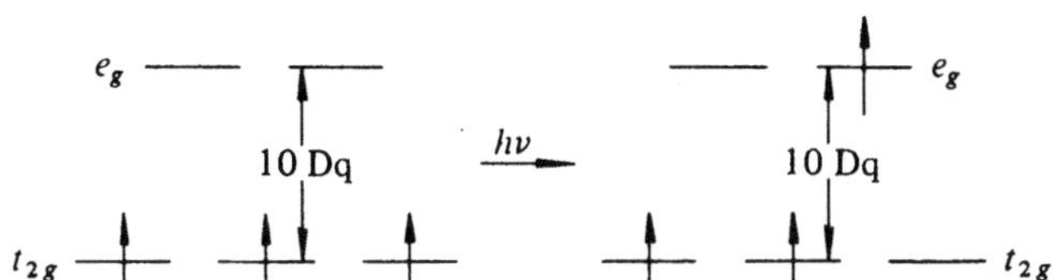

Scheme 2

the $^4T_{1g}$ and $^4T_{2g}$ (and the corresponding doublet states) states depends significantly on the ligands bound to Cr(III).

Both the excited states arising from the t_{2g}^3 configuration[12, 13] and quartets[7] from $t_{2g}^2e_g^1$ were invoked as possible reactive states. Definitive experimental evidence is now available which shows that the low lying doublet states are unreactive with respect to photosubstitution in $Cr(CN)_6$.[3–14] Reaction (4) occurs with 0.08 and 0.09 quantum efficiency upon 305 nm and 370 nm photolysis, respectively.[15] The approximate energetic positions of the low lying LF states of $Cr(CN)_6^{3-}$ are shown

$$Cr(CN)_6^{3-} \xrightarrow[\substack{pH = 6.8 \\ 25°, H_2O}]{h\nu} Cr(CN)_5(OH_2)^{2-} + CN^- \tag{4}$$

in Fig. 1.[14] As expected for the strong LF strength ligand, CN^-, the states arising from $t_{2g}^2e_g^1$ are much higher energetically than the t_{2g}^3 excited states and chemical spectroscopic techniques developed by organic photochemists can be applied. A

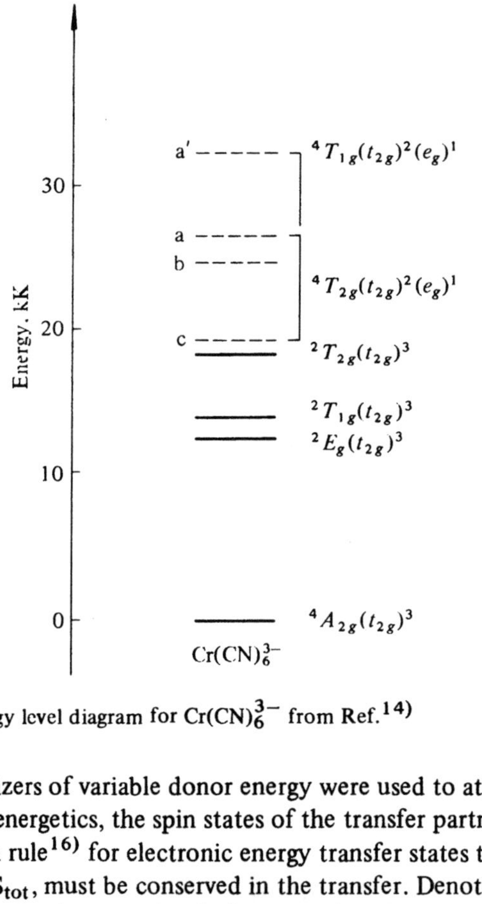

Fig. 1. Ligand field energy level diagram for $Cr(CN)_6^{3-}$ from Ref.[14]

series of triplet sensitizers of variable donor energy were used to attempt to sensitize reaction (1). Besides energetics, the spin states of the transfer partners are important. The spin-conservation rule[16] for electronic energy transfer states that the total spin angular momentum, S_{tot}, must be conserved in the transfer. Denoting the total spin angular momentum of the donor molecule (sensitizer) as S_D and that of the acceptor molecule $[Cr(CN)_6^{3-}]$ as S_A, S_{tot} has components $S_D + S_A, S_D + S_A - 1, S_D + S_A - 2, S_D + S_A - 3, \ldots, S_D - S_A$. For an interaction to be spin-allowed it is required that the $S_{tot\,(before)}$ and $S_{tot(after)}$ have least one component in common. Applying this to the case at hand we know that all of the donors are triplets so $S_{D(before)} = 1$ and are deactivated to singlets upon electronic energy transfer so $S_{D(after)} = 0$. The acceptor, $Cr(CN)_6^{3-}$, is a quartet so $S_{A(before)} = 3/2$ and can be excited to a doublet or a quartet, $S_{A(after)} = 1/2$ or $3/2$, respectively. These interactions are written out separately in Eqs. (5) and (6) for doublet and quartet production, respectively. Each

$$^{3}(\text{Donor}) + {}^{4}[\text{Cr(Cr)}_6^{3-}] \longrightarrow {}^{1}(\text{Donor}) + {}^{2}[\text{Cr(CN)}_6^{3-}] \quad (5)$$
$$S_D=1 \quad S_A=3/2 \qquad S_D=0 \quad S_A=1/2$$

S_{tot} = 5/2 (before) S_{tot} = 1/2 (after)
Components; 5/2, 3/2, 1/2 Components; 1/2

$$({}^{3}\text{Donor}) + {}^{4}[\text{Cr(CN)}_6^{3-}] \longrightarrow {}^{1}(\text{Donor}) + {}^{4}[\text{Cr(CN)}_6^{3-}] \quad (6)$$
$$S_D=1 \quad S_A=3/2 \qquad S_D=0 \quad S_A=3/2$$

S_{tot} = 5/2 (before) S_{tot} = 3/2 (after)
Components; 5/2, 3/2, 1/2 Components; 3/2

has one compenent in common; thus, both doublet excited state or quartet excited state production is spin-allowed using the triplet donors. It was found that sensitizers capable of exothermic transfer to ${}^4T_{2g}(t_{2g}^2e_g^1)$ do sensitize reaction (4). However, Ru(2,2′-bipyridine)$_3^{2+}$, which has too low a triplet energy to populate ${}^4T_{2g}(t_{2g}^2e_g^1)$, sensitizes the ${}^2E_g(t_{2g}^3) \rightarrow {}^4A_{2g}(t_{2g}^3)$ luminescence of $Cr(CN)_6^{3-}$, but no sensitization of the aquation was found. Thus, if one accepts that the high energy sensitizers actually produce the ${}^4T_{2g}(t_{2g}^2e_g^1)$ excited state (there is no direct evidence for this) then the reactive states are concluded to be those arising from the $t_{2g}^2e_g^1$ for $Cr(CN)_6^{3-}$. The experiments do show that the spin-pairing doublet states at low energy are not reactive for this case.[14)]

Another line of experimental evidence points to substitutional reactivity only for the $t_{2g}^2e_g^1$ excited states of $Cr(CN)_6^{3-}$. The ${}^2E_g(t_{2g}^3) \rightarrow {}^4A_{2g}(t_{2g}^3)$ emission of $Cr(CN)_6^{3-}$ can be totally quenched while the substitution reaction is unaffected[17)]. This experiment points to differing origins for the luminescence and the substitution, consistent with ${}^4T_{2g}(t_{2g}^2e_g^1)$ reactivity inferred from the Ru(2,2′-bipyridine)$_3^{2+}$ sensitization study outlined above.

These techniques of sensitization and quenching are not likely to be applicable to many other Cr(III) complexes. At least, the results will not be as definitive. The reasoning here rests with the electronic nature of the Cr(III) complexes where the LF strength is substantially smaller than that provided by the six CN^-'s. In many cases the ${}^4T_{2g}(t_{2g}^2e_g^1)$ and the doublet states from the t_{2g}^3 configuration are close enough energetically to allow thermal population of the ${}^4T_{2g}(t_{2g}^2e_g^1)$ state from, say, the ${}^2E_g(t_{2g}^3)$ state. Such a complication clouds the interpretation of sensitization and quenching studies. At the present time, though, there is no direct evidence that doublet states yield substitution and it is tempting to regard the results for $Cr(CN)_6^{3-}$ as general.

Adamson's rules do not apply to O_h systems, but both Zink's model[8)] and that proposed by Wrighton, Gray, and Hammond[5)] do. In both of these MO–LF models their proponents attempt to determine the bonding consequences upon one-electron transitions resulting from the absorption of light. Wrighton, Gray, and Hammond pointed out[5)] that quartet $(t_{2g}^3) \rightarrow$ doublet (t_{2g}^3) transitions result in little or no change in bonding, and thus the doublet states of t_{2g}^3 should not be substantially more reactive (or stable) than the ground state. The t_{2g} orbitals are of π symmetry and Zink does assign[9] some changes in π bonding upon spin pairing for some one-electron changes in tetragonal complexes but makes no definitive remarks for the O_h case. It can be inferred, however, from his discussion that some reactivity could be expected from the doublet states. Both of the models rightly predict enhanced substitutional reactivity for σ-donor or π-acceptor ligands with the $t_{2g}^3 \rightarrow t_{2g}^2 e_g^1$ type excitation. The e_g orbitals are strongly σ-antibonding and thus population of them will yield weakened metal-ligand bonding in a σ sense. At the same time depopulation of t_{2g} by one electron either does not affect the bonding (σ-donor ligands) or weakens π-bonding (π-acceptor ligands). For π-donor ligands, the t_{2g} level is weakly π-antibonding. Thus, for π-donor ligands the changes in bonding are compensating effects: weakened σ-bonding and strengthened π-bonding. Results for low-spin d^6 complexes (*vide infra*) support the notion that the chemical consequence of weakening in the σ-framework is greater than the strengthening in the π-framework, but this is not to say that the changes in π-bonding are inconsequential as will be shown below for non-O_h complexes. At least, the theories do provide a reasonable rationale for the substitution reactivity of the $^4T_{2g}(t_{2g}^2 e_g^1)$ state.

Complexes of Cr(III) of C_{4v} and D_{4h} symmetry have received a great deal of attention recently and some fairly important results have emerged. Data like that shown in Table 1 led Adamson to propose his empirical rules:[7] (1) labilization occurs along the axis of the octahedron having the smallest average ligand field and (2) if the labilized axis contains two different ligands, the one of greater ligand field strength is preferentially aquated.

Table 1. Early photoaquation data for non-O_h Cr(III) complexes[1])

Complex	Ligand Aquated	ϕ[2])
$Cr(NH_3)_5OH_2^{3+}$	NH_3	~0.15–0.20
$Cr(NH_3)_5(NCS)^{2+}$	NCS^-	~0.13–0.18
$Cr(NH_3)_5Cl^{2+}$	NH_3	~0.35–0.39
$t-Cr(NH_3)_2(NCS)_4^-$	NCS^-	~0.29–0.32
$Cr(OH_2)_5NCS^{2+}$	NCS^-	$2.1–6.0 \times 10^{-5}$

[1]) Ref.[7]).
[2]) Yields are for irradiation into lowest LF quartet → quartet absorptions.

It was not until 1971 when Pyke and Linck published[18] their study of *trans*-$Cr(en)_2F_2^+$ that any serious experimental exception to the rules could be made. The photochemistry of the ~D_{4h} *trans*-$Cr(en)_2F_2^+$ is indicated in reaction (7). The F-Cr-F axis is clearly the axis of lowest LF strength and yet F^- substitution does

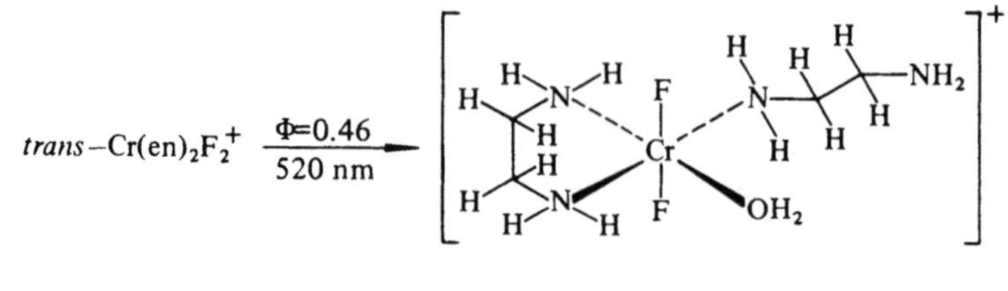

(7)

not exclusively obtain. Pyke and Linck reasoned that the lowest excited state features population of $d_{x^2-y^2}$ ccording to the one-electron level scheme expected in D_{4h}, Scheme 3. The $d_{x^2-y^2}$ orbital has σ-antibonding character for the ligands on the x- and y-axes and should give the result indicated in reaction (7) based on the "σ-bonding

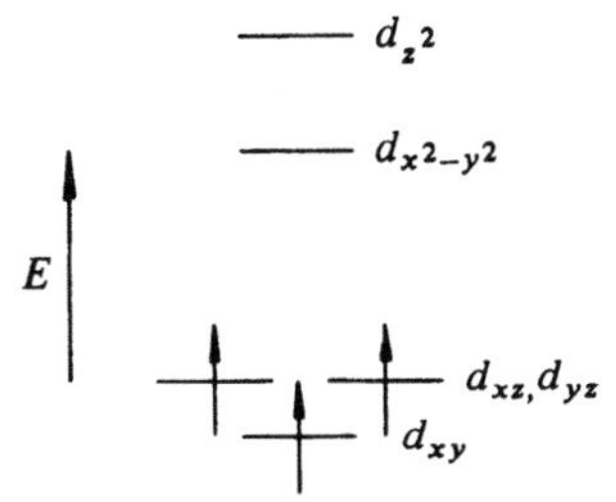

Scheme 3

approach". Actually, however, the lowest excited state cannot be viewed as a one-electron population of $d_{x^2-y^2}$ and indeed LF calculations show that the lowest excited state is no less than 66% d_{z^2} in character[5]. Thus, the σ-bonding approach does not account for the result indicated in reaction (7). Wrighton, Gray, and Hammond[5] first successfully rationalized the cleavage of the Cr-N bond by pointing out that in a one-electron sense the excited state is not substantially different than in O_h, *i.e.* though the symmetry is only D_{4h} the splitting of $d_{x^2-y^2}$ and d_{z^2} is fairly small as reflected in the LF calculations showing approximately 66/34 $d_{z^2}/d_{x^2-y^2}$ character in the lowest excited state. Thus, both the Cr-F and the Cr-N bond would be labilized in a σ-sense. However, depopulation of the d_{xz}, d_{yz}, d_{xy} set *increases* the π-bonding for the Cr-F bond, since F is a π-donor. No such compensating effect obtains for the Cr-N bond upon depopulation of the π-d orbitals, since the amine is only a σ-donor. These effects are summarized in Scheme 4 and it was concluded that the Cr-N bond will be labilized more so than the Cr-F bond *compared to the ground state.* From this example we learn that changes in π-bonding can play a key role in

$$[trans\text{-}Cr(en)_2F_2]^+ \xrightarrow{520\ nm} {}^*[trans\text{-}Cr(en)_2F_2]^+$$

$$(\pi\text{-}d)^3 \longrightarrow (\pi\text{-}d)^2(\sigma^*)^1$$

Bond	Change in Bonding
Cr-F	σ-weakened π-strengthened
Cr-N	σ-weakened

Scheme 4

the stereochemistry of photosubstitution. Wrighton, Gray, and Hammond emphasized the fact that their predictions reflect the relative substitution rate constants of the excited state in question compared to the ground state with the assumption that the mechanism for substitution in each state is principally dissociative in nature.[5]

By way of contrast, Zink's model[8, 9] is not based on the assumption of a dissociative type mechanism.

We can explain the empirical rules proposed by Adamson[7] using the language and models outlined above. One example will be illustrated. Consider $Cr(NH_3)_5Cl^{2+}$ which undergoes the reactions (8) and (9) with the indicated quantum yields.[19] The selective loss of NH_3 to yield principally the *cis*-$Cr(NH_3)_4(OH_2)Cl^{2+}$ product is consistent with none of the models unless the NH_3 which is lost comes from a position *trans* to the Cl^- in $Cr(NH_3)_5Cl^{2+}$. The photolysis of *cis*- and *trans*-

$$Cr(NH_3)_5Cl^{2+} \xrightarrow[\Phi = 0.35-0.4]{h\nu,\ \text{LF region}} cis\text{-}Cr(NH_3)_4(OH_2)Cl^{2+} + NH_3 \quad (8)$$

$$Cr(NH_3)_5Cl^{2+} \xrightarrow[\Phi = 0.007-0.05]{h\nu,\ \text{LF region}} Cr(NH_3)_5(OH_2)^{3+} + Cl^- \quad (9)$$

$Cr(NH_3)_4{}^{15}NH_3Cl^{2+}$ in the LF region reveals that greater than 75% of the NH_3 lost does come from the position *trans* to the Cl^-.[20] Adamson's rules[7] are thus obeyed, but the facts are best interpreted using MO and LF arguments. For $Cr(NH_3)_5Cl^{2+}$ the population of the d_{z^2} orbital is the essential result of absorption of light corresponding to the lowest LF excited state. Thus, the H_3N-Cr-Cl axis is labilized in a σ-sense, but only the Cr-Cl bond is strengthened in a π-sense by depopulation of the π-d orbitals. The Cr-NH_3 bond, then, is most labilized consistent with the experiments mentioned above. Again the role of π-bonding is key to the stereochemical course of the photosubstitution.

Thus far, consideration has only been given the reactivity of the lowest LF state arising from a π-$d \rightarrow \sigma^*$ type excitation. In a complex such as $Cr(NH_3)_5Cl^{2+}$ the direct population of a higher LF state may result in the population of the $\sigma^*_{xy}(d_{x^2-y^2})$ orbital rather than the $\sigma^*_z(d_{z^2})$ orbital. In such a case loss of an equatorial NH_3 would be expected if such a process can compete with internal conversion to the lowest LF state which would populate the $\sigma^*_z(d_{z^2})$. However in a series of $Cr(NH_3)_5X^{n+}$ complexes the reaction and quantum yields are independent of which one-electron excitation is achieved.[21] This result implicates fast internal conversion and provides little information concerning the reactivity of the upper LF states.

One other type of reactive excited state merits discussion here. Photolysis in the region of ligand → Cr CT absorption often leads to a substantial quantum yield for net substitution of the ligand involved in the CT. Some substantiating data are found in Table 2 for two compounds.[21] Current thought focusses on consideration of the CT state as a sort of LF excited state of Cr(II) which is very thermally labile.[20] For a species such as $Cr(NH_3)_5\ Cl^{2+}$ which yields fairly high quantum yields for net Cl^- substitution there is also evidence that the NH_3 groups are labilized in the Cl → Cr CT excited state. Such a result is not particularly surprising, but since the LF states also yield high quantum yields for NH_3 substitution, the extent of NH_3 substitutional reactivity of the CT state is not clear. In this regard one crucial measurement which is lacking is the CT → LF internal conversion efficiency. The data do, however, justify the statement that the CT state is reactive with respect to net substitution without final reduction of Cr(III) to Cr(II).

Table 2. Net photoaquation in L → Cr(III)CT absorption[1])

Complex	Irrdn λ, nm	Type of reaction	ϕ
$Cr(NH_3)_5Cl^{2+}$	250	NH_3aquat.	0.35
		Cl^- aquation	0.23
$Cr(NH_3)_5Br^{2+}$	250	NH_3 aquat.	0.20
		Br^- aquat.	0.26

[1]) Ref.[21]).

2. The Mechanism of Photosubstitution in Cr(III) Complexes

Discussion of the mechanism of photosubstitution in Cr(III) complexes has largely concerned the possibility of a seven-coordinate intermediate. Such an associative type mechanism is not proven by any means, but there is some circumstantial data to support it and some data that tends to rule out a purely dissociative mechanism involving the photoinduced formation of a discrete five-coordinate intermediate capable of being competitively scavenged by nucleophiles in the medium.

The first piece of pertinent information is that for several $Cr(NH_3)_5X^{2+}$ complexes the stereochemistry of the product is *cis*-$Cr(NH_3)_4XY^{2+}$ where Y is the entering group. Some of the complexes investigated and their photosubstitution products are given in Table 3.[19, 22–28]) If we take these results at face value it is clear that Y does not occupy the binding site occupied by the leaving NH_3 group.

Table 3. Stereochemistry of photoaquation in Cr(III) complexes

Complex	Product	ϕ	Ref.
$Cr(NH_3)_5Cl^{2+}$	*cis*-$Cr(NH_3)_4(OH_2)Cl^{2+}$	0.36	19)
$Cr(NH_3)_5Br^{2+}$	*cis*-$Cr(NH_3)_4(OH_2)Br^{2+}$	0.35	22)
$Cr(NH_3)_5(CF_3COO)^{2+}$	*cis*-$Cr(NH_3)_4(OH_2)(CF_3COO)^{2+}$	0.36	23)
$Cr(NH_3)_5(NCS)^{2+}$	*cis*-$Cr(NH_3)_4(OH_2)(NCS)^{2+}$	0.42	24, 25)
trans-$Cr(en)_2Cl_2^+$	*cis*-$Cr(en)_2(OH_2)Cl^{2+}$(> 70%)	0.31	26)
trans-$Cr(NH_3)_4Cl_2^+$	*cis*-$Cr(NH_3)_4(OH_2)Cl^{2+}$	0.37	27)
trans-Cr(cyclam)Cl_2^+	*trans*-Cr(cyclam)$(OH_2)Cl^{2+}$	$\sim 4 \times 10^{-4}$	28)
trans-Cr(2, 3, 2 tet)Cl_2^+	*cis*-Cr(2, 3, 2 tet)$(OH_2)Cl^{2+}$	0.06	28)

Rather, the data are consistent with the statement that Y occupies a position *cis* to the ligand which was *trans* to the leaving group in the original complex. This result seemingly cannot be accomodated by the generation of a 5-coordinate intermediate as it would yield *trans*-$Cr(NH_3)XY^{2+}$ if the intermediate is a rigid C_{4v} square-pyramid or a mixture of *cis*- and *trans*-$Cr(NH_3)_4XY^{2+}$ if the intermediate is a D_{3h} trigonal-bipyramid. Likewise, the photoinertness of *trans*-Cr(cyclam)Cl_2^+ compared to *trans*-$Cr(en)_2Cl_2^+$ tends to rule out a photodissociation of Cl^- as the mechanism for Cl^- substitution. The stereochemically constraining cyclam ligand presumably prevents the Cr(III) complex from adopting a conformation capable of developing binding interactions with the entering group. One other experimental result lends credence to a mechanism which is not truly dissociative. The substitution quantum yield for several Cr(III) complexes has been shown to depend on the medium.[29] While medium effects on Cr(III) excited states may simply change their lifetime, one could interpret the medium effects as those expected when an associative mechanism for substitution obtains. A definitive investigation of entering group effects has not been reported as virtually every report of Cr(III) photosubstitution is really a photoaquation reaction.

One mechanistic point to be emphasized is that in numerous cases there is a real difference between the stereochemical course of the photosubstitution and the thermal substitution. In the cases where this fact does pertain we thus have evidence against conversion of electronically excited states to the intermediate or transition state involved in the ground state reaction. This aspect of the mechanism leads to the notion that the energy difference between the excited state and the ground state primary product (or intermediate) becomes smaller as the excited state distorts toward the structure of the ground state product. This sort of Hammond postulate for excited to ground state conversions has value in that as the excited state structurally and energetically approaches the ground state product, then nonradiative decay to that product will be fast since the electronic relaxation becomes vertical and involves a small energy gap. By applying such a postulate to Cr(III) complexes one rationalizes that a different, perhaps more reactive and energetic intermediate, can be photogenerated than that involved in thermally activated substitution. This line of reasoning suggests that the nature of the excited state distortion and the energetic proximity to a ground state intermediate or product will be key considerations in prediction of the course of the reaction.

B. Photosubstitution in Low-Spin d^6 Complexes

In contrast to the work on the d^3 Cr(III) system investigation of d^6 complexes will be generally useful as there are many metal complexes of this electronic configuration. To illustrate, some complexes of the metals or ions included in Table 4 having

Table 4. Central metal of low-spin d^6 complexes which undergo photosubstitution

Group VB	Group VIB	Group VIIB	Group VIII		
V(–I)	Cr(0)	Mn(I)	Fe(II)	Co(III)	
Nb(–I)	Mo(0)		Ru(II)	Rh(III)	
Ta(–I)	W(0)	Re(I)	Os(II)	Ir(III)	Pt(IV)

low-spin d^6 configurations have been the object of some photosubstitution study. In this section we begin first by a consideration of LF states for low-spin d^6 followed by a discussion of the Group VIII d^6 complexes and finally Group VB, VIB, and VIIB will be discussed together. The reason for grouping Group VIII systems as a unit is that these complexes involve higher oxidation states, II, III, and IV, and the systems most well-studied are aqueous systems. Group VB, VIB, and VIIB d^6 complexes with low valent metals, –I, 0, and I are largely organometallic systems. The organometallic systems have been generally studied in nonaqueous media.

1. LF States of Low-Spin d^6 Complexes

Like the d^3 Cr(III) systems much spectroscopic work has been carried out on six-coordinate low-spin d^6 complexes, and it can now be said that the essential features of the LF spectra are understood.[3, 30–42] However, unlike the Cr(III) systems, the low-spin d^6 systems are not well understood with respect to their luminescense phenomena. This is due, in part at least, to the diversity of luminescent complexes

Table 5. Spin allowed ligand field electronic absorption bands for O_h low-spin d^6 complexes

Complex	${}^1A_{1g} \rightarrow {}^1T_{1g}(\epsilon)$, kK	${}^1A_{1g} \rightarrow {}^1T_{2g}(\epsilon)$ kK	Ref.
$Co(OH_2)_6^{3+}$	16.5	24.7	30)
$Co(NH_3)_6^{3+}$	21.2(56)	29.5 (46)	31)
$Co(en)_3^{3+}$	21.5 (88)	29.6 (78)	31)
$Co(CN)_6^{3-}$	32.4 (200)	39.0 (140)	31)
$Rh(OH_2)_6^{3+}$	25.5 (47)	32.8 (55)	32)
$Rh(NH_3)_6^{3+}$	32.8 (134)	39.2 (101)	33)
$Rh(en)_3^{3+}$	33.2 (210)	39.6 (190)	31)
$Rh(CN)_6^{3-}$	44.4 (555) sh	–	34)
$Ir(NH_3)_6^{3+}$	39.8 (92)	46.8 (160)	35)
$Ir(CN)_6^{3-}$	> 52.0	–	34)
$PtCL_6^{2-}$	28.3 (490)	–	36)
PtF_6^{2-}	31.5	36.4	37)
$Fe(CN)_6^{4-}$	31.0 (~ 300)	37.0 sh	38)
$Ru(OH_2)_6^{2+}$	18.9 (21)	25.6 (31)	39)
$Ru(en)_3^{2+}$	27.0(40)	33.2(91)	40)
$Mn(CO)_6^+$	~38.0 (~1500)	–	41)
$Re(CO)_6^+$	~39.0 (~2000)	47.1 (4600)	41)
$Cr(CO)_6$	~30.0 (~2000)	38.85 (3500)	41)
$Mo(CO)_6$	~31.0 (~2200)	37.20 (7900)	41)
$W(CO)_6$	~31.0 (~2400)	37.10 (7400)	41)
$V(CO)_6^-$	~24.0 (~1000)	~31.1 (3300)	41)
$Nb(CO)_6^-$	~24.0 (~1000)	–	42)
$Ta(CO)_6^-$	~24.0 (~1000)	–	42)

and lowest excited states involved. Along with their thermal stability, it is just this diversity, though, that will make the low-spin d^6 systems the most well studied, understood, interesting, and perhaps, usefui inorganic photochemical systems.

The ground state $^1A_{1g}$ of low-spin d^6, O_h systems has the t_{2g}^6 electronic configuration and the fully filled t_{2g} level endows these complexes with special thermal stability. The lowest LF excited states achieved by a one-electron transition are all of the $t_{2g}^5e_g^1$ electronic configuration and give rise to the $^3T_{1g}$, $^3T_{2g}$, $^1T_{1g}$, and $^1T_{2g}$ states. Generally, these states are the only ones considered; little importance has been given the $^5T_{2g}(t_{2g}^4e_g^2)$ state which crosses below the $^3T_{1g}$ as the LF strength is diminished. The two-electron transition, $^1A_{1g}(t_{2g}^6) \rightarrow {}^5T_{2g}(t_{2g}^4e_g^2)$ is not likely to have a high absorption probability, but the population of the $^5T_{2g}(t_{2g}^4e_g^2)$ may occur by the intersystem crossing sequence indicated in Scheme 5 where the distortion indicated retains O_h symmetry. Generally, even the spin-forbidden $^1A_{1g}(t_{2g}^6) \rightarrow {}^3T_{1g}$, $^3T_{2g}(t_{2g}^5e_g^1)$ transitions are not observed except for the heavy metal third row systems. The absorption bands most widely reported are the spin-allowed $^1A_{1g}(t_{2g}^6) \rightarrow {}^1T_{1g}$, $^1T_{2g}(t_{2g}^5e_g^1)$ bands. These band positions are collected in Table 5 for a number of O_h, low-spin d^6 systems.[30–42] The energetic separation of the $^1T_{1g}(t_{2g}^5e_g^1)$ and $^1T_{2g}(t_{2g}^5e_g^1)$ states is seen to be fairly independent of the particular central metal as predicted from the Tanabe-Sugano diagram for the d^6 electronic configuration. As usual, though, the value of 10Dq depends on both ligand and metal. One final note about the data in Table 5 concerns the band inten-

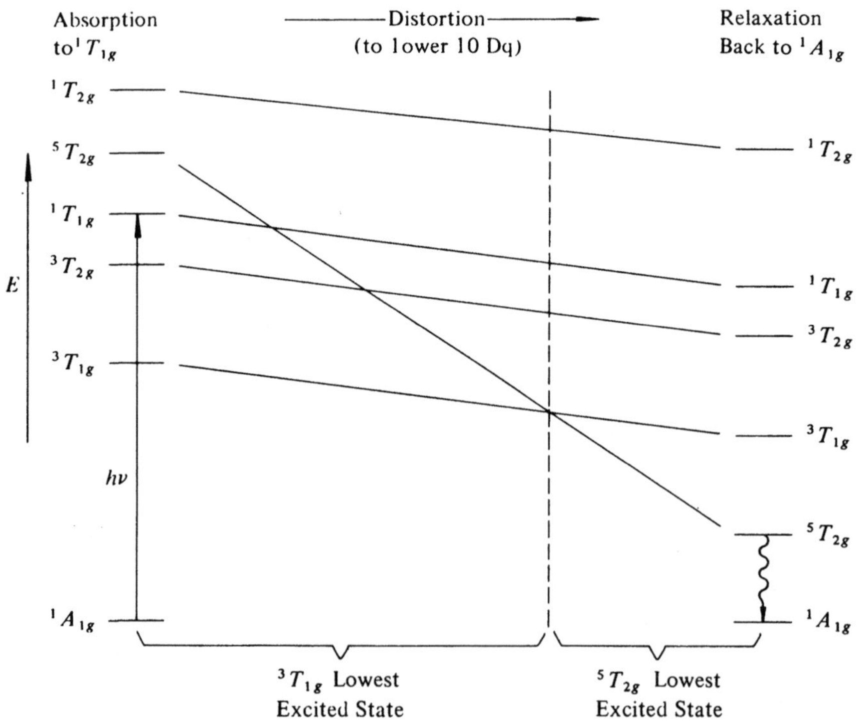

Scheme 5

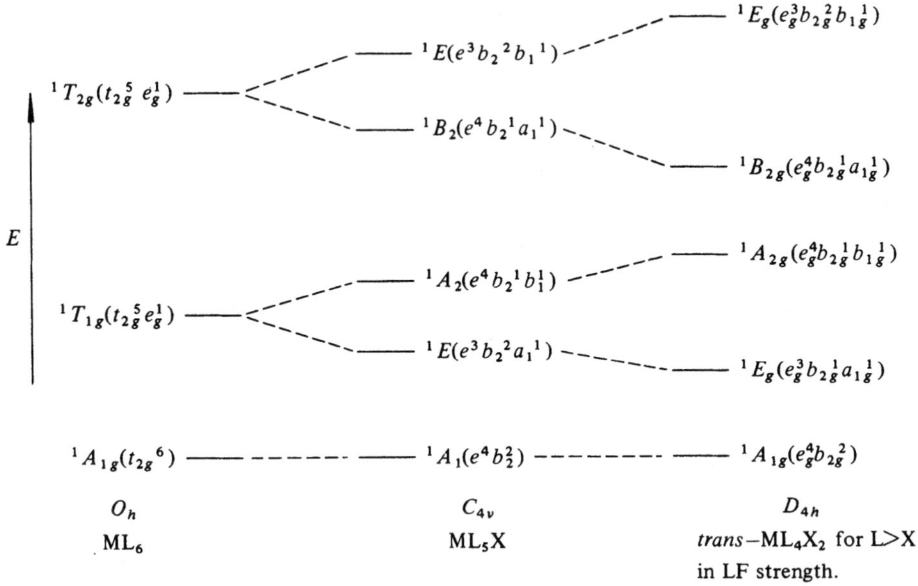

Scheme 6

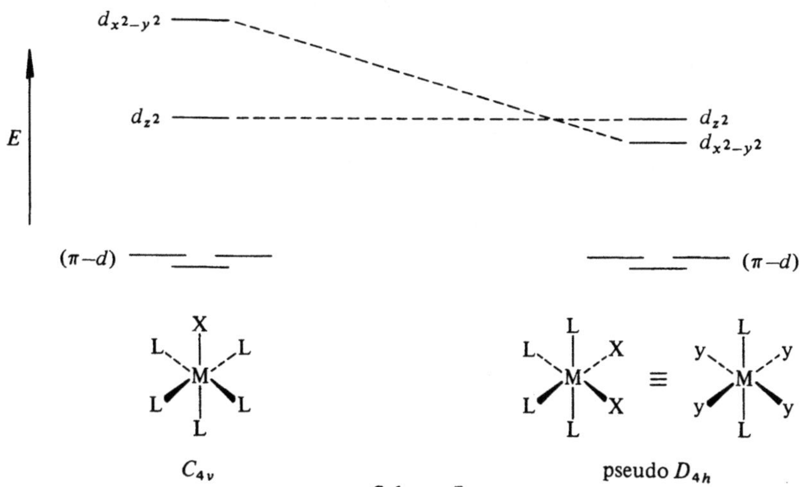

Scheme 7

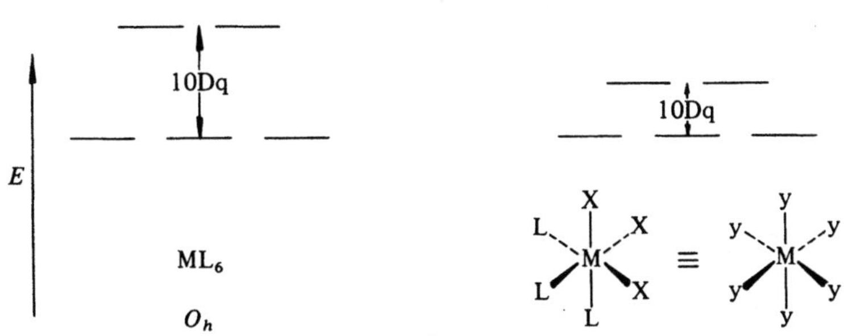

Scheme 8

sities. The more covalent metal carbonyls generally have the highest molar absorptivities and these are seen to be as much as several thousand $1\,mol^{-1}\,cm^{-1}$. Speaking within the framework of the models[5, 8, 9)] for excited state substitutional reactivity, the $t_{2g}^5 e_g^1$ configuration should yield excited states having weaker σ-bonding than the t_{2g}^6 ground state due to population of the $(e_g\sigma^*)$ orbital, and either a weakened, unaffected, or strengthened π-bonding compared to the ground state depending on whether the six-equivalent ligands are π-acceptor, σ-donor, or π-donor, respectively.

The non-O_h low-spin d^6 complexes are very important with respect to discussion of photosubstitution, and electronically they are reasonably well understood. In fact, it was with C_{4v}, low-spin d^6 complexes that the spectrochemical series of ligands was first formulated.[43)] Thus far the complexes of C_{4v}, D_{4h}, and C_{2v} symmetry that have been of most interest are those where the X in ML_5X, *trans*-ML_4X_2, or *cis*-ML_4X_2 is of lower LF strength than L. For the series $O_h \rightarrow C_{4v} \rightarrow D_{4h}$ the lowest LF singlet states change as indicated in Scheme 6.[44)] The one-electron configurations are given for each state and these can be used to assess the changes in bonding from state to state. Despite its C_{2v} symmetry, the *cis*-ML_4X_2 case can be treated in LF terms in a way that leads to clear cut predictions concerning changes in bonding due to the one-electron transitions. We treat the *cis*-ML_4X_2 case as *trans*-ML_2Y_4 where the LF strength of L is greater than Y and the LF strength of Y is equal to the average LF strength of L and X.[45)] Thus, in *cis*-ML_4X_2 there are two equivalent L-M-X axes and the problem becomes one of D_{4h} symmetry. The one-electron ordering in ML_5X and *cis*-ML_4X_2 are given in Scheme 7. In the same way that this *cis*-ML_4X_2 can be treated as a pseudo D_{4h} problem we can treat *fac*-ML_3X_3 as a pseudo O_h system as shown in Scheme 8. As all axes in *fac*-ML_3X_3 complexes are equivalent as they are in actual O_h complexes, these complexes may provide the best tests of theories predicting relative labilities of the ligands L and X. In any case the photosubstitution behavior of complexes below O_h symmetry provide stereochemical information which may be correlated with the one-electron diagrams given above. Some organmetallic complexes represent unique electronic situations and will be briefly discussed when dealing with their photosubstitution behavior (*vida infra*).

2. Photosubstitution in d^6 Group VIII Complexes

Of the Group VIII systems which have been studied Co(III) complexes are the best understood photochemically and electronically. Generally, Co(III) – cyanide and Co(III) – ammine complexes have been most well-studied and the former are the most photosensitive;[46)] but the latter have recently been examined.[47)] With respect to photosubstitution though the quantum yields for reaction are very small ($< 10^{-2}$) upon population of the LF excited states. The O_h $Co(CN)_6^{3-}$ complex undergoes reaction (10) with a quantum yield of ~0.3 at either 254 or 313 nm, *i.e.*, direct population of either the ${}^1T_{1g}$ or the ${}^1T_{2g}$ state yields substitution with the same quantum ef-

$$Co(CN)_6^{3-} \xrightarrow{h\nu} Co(CN)_5(OH_2)^{2-} + CN^- \qquad (10)$$

ficiency.[48)] Biacetyl triplet sensitized aquation of $Co(CN)_6^{3-}$ implicates ${}^3T_{1g}(t_{2g}^5 e_g^1)$ as a precursor to the aquation indicated in reaction (10).[49)] Such consideration does seem appropriate, but if we examine the Tanabe-Sugano diagram for O_h d^6 systems, and using the band positions given in Table 5, we find that $B \approx 450$ cm^{-1} and if the 0–0 emission band[50] for $Co(CN)_6^{3-}$ falls at ~17,000 cm^{-1} this yields an E/B value of ~37. At this value of E/B in the Scheme 5 we are still well within the region where the $^3T_{1g}$ is the lowest excited state. However, the lowest portion of the emission band is ~10,000 cm^{-1} and at this value for E one is close to the $^3T_{1g} \rightarrow {}^5T_{2g}$ crossover and it appears reasonable, on energetic grounds, to invoke such a crossover at least in aqueous solution at room temperature. The $^5T_{2g}(t_{2g}^4e_g^2)$ features two electrons in the σ^* e_g orbitals and should, therefore, be very substitution labile. For $Co(NH_3)_6^{3+}$ the E/B value for the $^3T_{1g}$ state is estimated to be ~26 from *absorption band maxima* for the $^1A_{1g} \rightarrow {}^1T_{1g}$ and $^1T_{2g}$ transitions. If E/B drops precipitously due to the symmetrical expansion upon $t_{2g}^6 \rightarrow t_{2g}^5e_g^1$ as in $Co(CN)_6^{3-}$ the $^5T_{2g}$ state should clearly be the lowest excited state in the system. The absence of emission and the lack of photosensitivity seem to point to a different situation in comparison to $Co(CN)_6^{3-}$. The lack of efficient LF photosubstitution generally found in Co(III) – ammines also contrasts to the situation with the Rh(III) – ammines (vide infra). Like Co(III) – cyanides, the Rh(III)-ammines represent a situation where the LF strength is large and the triplet, not the quintet, is likely the lowest spectroscopic state.

In all of the low-spin d^6 systems the one-electron LF transitions must populate either the d_{z^2} or the $d_{x^2-y^2}$ orbital and there is compelling data for the non-O_h systems to associate some stereochemical significance to the one-electron configuration of the lowest excited states. First, consider the C_{4v} $Co(CN)_5X^{3-}$ systems which are included in Table 6.[48, 49, 51–54] In many cases $X << CN^-$ in LF strength, and Scheme 6 is applicable.[55] The photosubstitution generally involves X, *i.e.*, the ligand on the *z*-axis which is labilized by population of the $d_{z^2}(a_1)\sigma_z^*$ orbital. Thus, the chemistry and one-electron orbital configurations seem consistent. Wavelength dependent photosubstitution reactions have been reported[53] for $Co(CN)_5(OH)^{3-}$ and $Co(CN)_5(OH_2)^{2-}$ where it was found that increasing yields for CN^- substitution obtain upon increasing the energy of the excitation light. This increase in CN^- substitution was ascribed to population of the $d_{x^2-y^2}(b_1)\sigma_{xy}^*$ orbital at the higher energy wavelengths. The larger yields for CN^- substitution in the hydroxo as compared to the aquo complex are rationalized by the fact that OH^- can more fully exploit π-bonding upon depopulation of the π-d orbitals and, hence, there is a compensation of the lability of the Co-O bond induced by population of d_{z^2}. For OH_2, though, which is principally a σ-donor there is no such compensating effect and labilization of the Co–OH_2 bond is greater than of the Co–OH bond. The slower rates for OH^- substitution in the excited $^{1,3}E(e^3b_2^2a_1^1)$ states inferred from these arguments allow CN^- substitution rates to be more competitive. The models[5, 8, 9] predict axial-CN^- release and at low temperature *trans*-$Co(CN)_4(OH)_2^{3-}$ is the principal product.[56]

Not only the $Co(CN)_5X^{3-}$, but also *trans*-$Co(CN)_4(SO_3)(Y)^{n-}$ ($Y = OH_2$, OH^-, SO_3^{2-}) complexes, seem to give photoreactivity patterns consistent with the fact that the d_{z^2} orbital is populated in the lowest LF state.[54] The photoreaction of *trans*-$Co(CN)_4(SO_3)(OH_2)^{3-}$ is particularly noteworthy in that SO_3^{2-} undergoes efficient photosubstitution, while it is inert to substitution thermally.[57] In fact,

Table 6. Photoaquation of cyanocobaltate(III) complexes

Complex	Irrdn λ, nm	Product	ϕ	Ref.
$Co(CN)_6^{3-}$	254	$Co(CN)_5OH_2^{2-}$	0.31	48)
	313	$Co(CN)_5OH_2^{2-}$	0.31	48)
	365	$Co(CN)_5OH_2^{2-}$	0.31	48)
$Co(CN)_5Cl^{3-}$	370	$Co(CN)_5OH_2^{2-}$	~0.06	51, 48)
$Co(CN)_5Br^{3-}$	370	$Co(CN)_5OH_2^{2-}$	~0.17	51, 48)
$Co(CN)_5I^{3-}$	380	$Co(CN)_5OH_2^{2-}$	0.17	52)
	500	$Co(CN)_5OH_2^{2-}$	0.17	52)
	550	$Co(CN)_5OH_2^{2-}$	~0.16	51, 48)
$Co(CN)_5OH_2^{2-}$	313	$Co(CN)_4(OH_2)_2^-$	0.003	53)
	366	$Co(CN)_4(OH_2)_2^-$	0.001	53)
$Co(CN)_5OH^{3-}$	313	$Co(CN)_4(OH)_2^{3-}$	0.06	53)
	366	$Co(CN)_4(OH)_2^{3-}$	0.04_7	53)
	405	$Co(CN)_4(OH)_2^{3-}$	0.02_4	53)
	436	$Co(CN)_4(OH)_2^{3-}$	0.02_2	53)
$Co(CN)_5(SCN)^{3-}$	436	$Co(CN)_5(OH_2)^{2-}$	0.18	49)
$Co(CN)_5N_3^{3-}$	436	$Co(CN)_5(OH_2)^{2-}$	0.22	49)
t-$Co(CN)_4(SO_3)_2^{5-}$	313	t-$Co(CN)_4(SO_3)(OH)^{4-}$	0.36	54)
	366	t-$Co(CN)_4(SO_3)(OH)^{4-}$	0.57	54)
	436	t-$Co(CN)_4(SO_3)(OH)^{4-}$	0.57	54)
t-$Co(CN)_4(SO_3)(OH_2)^{3-}$	254	$Co(CN)_4(OH_2)_2^-$	0.15	54)
	313	$Co(CN)_4(OH_2)_2^-$	0.11	54)
	366	$Co(CN)_4(OH_2)_2^-$	0.14	54)
t-$Co(CN)_4(SO_3)(OH)^{4-}$	254	$Co(CN)_4(OH)_2^{3-}$	0.13	54)
	313	$Co(CN)_4(OH)_2^{3-}$	0.09	54)
	366	$Co(CN)_4(OH)_2^{3-}$	0.19	54)

SO_3^{2-} generally has a large *trans* effect associated with it and this complex is no exception in that the OH_2 group is very thermally labile. The lability in the lowest excited state of SO_3^{2-} is ascribed to its role as a π-acceptor being labilized both by depopulation of the $d_{xz}, d_{yz}(e)$ (π-d) orbitals *and* by population of the $d_{z^2}(a_1)\sigma_z^*$ orbital. The loss of SO_3^{2-} from *trans*-$Co(CN)_4(SO_3)(OH_2)^{3-}$ and CN^- from $Co(CN)_5(OH_2)^{2-}$ seem to represent the only two known examples of Co(III)-cyanide photochemistry where a difference obtains in relative ground state rates and relative excited state rates for ligand substitution.

Turning now to other non-O_h complexes where a correlation may exist between one-electron excited state configurations and the stereochemistry of photosubstitution we can consider the Rh(III) and Ir(III) homologues of Co(III)-ammines and Co(III)-cyanides. Literature data for $Rh(CN)_5(X)^{3-}$ and $Ir(CN)_5(X)^{3-}$ is meager but do show, at least, that for $X << CN^-$ in LF strength that X undergoes substitution upon excitation into the $^1E(e^3b_2^2a_1^1)$ state. The substantiating data are given in Table 7.[34] The absorption maximum for the lowest $^1A_1 \rightarrow {}^1E$ varies as expected from the spectrochemical series for X, and both the absorption data and photochemistry parallel that for the Co(III) analogues.

Table 7. Photoaquation of Rh(III) and Ir(III) cyanide complexes [1]

Complex	$^1A_1 \rightarrow {}^1E$(kK)[2]	Product[3]
$Rh(CN)_5(XCCH_3)^{2-}$	40.8	$Rh(CN)_5(OH_2)^{2-}$
$Rh(CN)_5Cl^{3-}$	36.1	$Rh(CN)_5(OH_2)^{2-}$
$Rh(CN)_5Br^{3-}$	34.8	$Rh(CN)_5(OH_2)^{2-}$
$Rh(CN)_5I^{3-}$	31.9	$Rh(CN)_5(OH_2)^{2-}$
$Ir(CN)_5(NCCH_3)^{2-}$	46.5	$Ir(CN)_5(OH_2)^{2-}$
$Ir(CN)_5Cl^{3-}$	40.8	$Ir(CN)_5(OH_2)^{2-}$
$Ir(CN)_5Br^{3-}$	38.9	$Ir(CN)_5(OH_2)^{2-}$
$Ir(CN)_5I^{3-}$	36.4	$Ir(CN)_5(OH_2)^{2-}$

[1]) Ref.[34].
[2]) pH = 6 aqueous solutions at 300 K.
[3]) Photolysis at 254 nm in pH = 6 aqueous solution at 300 K.

The Rh(III)- and Ir(III)-ammines undergo efficient photosubstitution in marked contrast to the Co(III) analogues as mentioned above. However, the efficient photosubstitution found for the second and third row metal systems does appear to obey the essential rationalizations outlined in the models for LF substitutional reactivity. Two sets of complexes seem to behave quite nicely in this regard: C_{4v}, $Rh(NH_3)_5X^{2+}$ [58] and D_{4h} *trans*-$M(en)_2X_2^+$ (M = Ir, Rh; X = Cl, Br, I).[59, 60] In both sets of complexes the lowest excited state is likely one which features population of the $d_{z^2}(\sigma_z^*)$ orbital. Consistent with this fact, the photosubstitution chemistry can be viewed as resulting from loss of a ligand on the *z*-axis. For the Rh(III)-amines a dissociative mechanism is suggested for Cl^- photoaquation by the quantum yield data in Table 8.[60] The

Table 8. Photoaquation quantum yields for several Rh(III)-amines[1])

Compound	Irrdn λ, nm	ϕ Cl^-	Product
$Rh(NH_3)_5Cl^{2+}$	358	0.13 ± 0.01	$Rh(NH_3)_5OH_2^{3+}$
t-$Rh(NH_3)_4Cl_2^+$	407	0.14 ± 0.01	*t*-$Rh(NH_3)_4(OH_2)Cl^{2+}$
	358	0.17	
t-$Rh(en)_2Cl_2^+$	407	0.057 ± 0.002	*t*-$Rh(en)_2(OH_2)Cl^{2+}$
t-$Rh(cyclam)Cl_2^+$	407	0.011 ± 0.001	*t*-$Rh(cyclam)(OH_2)Cl^{2+}$

[1]) Ref.[60].

retention of stereochemistry upon Cl^- substitution and the small effect on reaction efficiency by the increased chelation supports a dissociative mechanism, and these data support the proposition that the photosubstitution mechanism is different compared to Cr(III) systems. Additional quantitative data for $Rh(NH_3)_5X^{2+}$ complexes are given in Table 9.[58a] These data show that as X is varied from Cl to Br to

Table 9. Photoaquation quantum yields for $Rh(NH_3)_5X^{2+}$ complexes at 25 °C[1])

X	Irrdn λ, nm	ϕX^-	ϕNH_3
Cl	350	0.16 ± 0.01	$< 10^{-3}$
Br	360	0.019 ± 0.001	0.18 ± 0.02
I	385	0.01	0.82 ± 0.08

1) Ref.[58a]).

the ligand undergoing substitution, X vs NH_3 [presumably axial since product is *trans*-$Rh(NH_3)_4(OH_2)X^{2+}$], changes from almost exclusively X for X = Cl to nearly all NH_3 at X = I. The axis of labilization is still the *z*-axis, but the decreasing X vs. NH_3 labilization can be rationalized as due to increasing π-donor interaction going down the halide group. Interestingly, biacetyl triplet sensitization seems to give nearly the same reaction quantum yields[58a]) for the $Rh(NH_3)_5X^{2+}$ complexes which suggest reaction occurs via the lowest ^3E LF state.

Some photoaquation studies of *bis*(2,2′-bipyridine), *bis*(1,10-phenanthroline) and *tetrakis*(pyridine) complexes of Rh(III) and Ir(III) complexes have also been carried out, and quantum yield data are given in Table 10.[61]) Except for the pyridine complexes, halide substitution is the only detectable photoreaction. The reactions likely occur from low lying LF states, but some information concerning excited state reactivity can be realized here. First, in the *cis*-$M(N_4)X_2^+$ (N = nitrogen donor) the X-M-N axes are labilized by the lowest one-electron transitions, Scheme 7. The X-M-N axis is also the axis of labilization in $Rh(NH_3)_5X$. The data in Table 10 for the *cis*-$M(N_4)X_2^+$ complexes show, somewhat convincingly, that I^- undergoes more efficient photosubstitution than either Br^- or Cl^-, which is the opposite trend to that in $Rh(NH_3)_5X^{2+}$ complexes. Another distinction is found in the *trans*-*tetrakis*(pyridine) complexes which undergo pyridine substitution with yields comparable to halide substitution. This is in contrast to *trans*-$M(en)_2X_2^+$ complexes where only halide substitution is found. The lability of pyridine is curious as we do not expect pyridine to be a good π-acceptor in these high oxidation state Rh(III) and Ir(III) complexes and consequently little enhanced lability of pyridine can be expected by depopulation of π-*d* orbitals. A detailed assessment of these facts must await thermal rate studies for M-N vs M-X cleavage in each case.

Studies of the photochemistry of Co(III)-amines have been recently[47, 62, 63]) carried out and do show that the complexes are somewhat photosensitive with respect to ligand substitution in the LF region. Several important facts have emerged. First,

Table 10. Photoaquation of Rh(III) and Ir(III) pyridyl, 2,2′-bipyridyl, and 1,10-phenanthroline halide complexes[1])

Complex	$\phi(X^-)$ 436 nm	366 nm	350 nm	334 nm	313 nm	254 nm
cis-$Rh(bipy)_2Cl_2^+$	0.032	0.035	–	–	–	0.013
cis-$Ir(bipy)_2Cl_2^+$	–	–	0.027–0.0486	–	–	0.02
cis-$Rh(bipy)_2Br_2^+$	0.035	0.037	–	–	–	0.021
cis-$Ir(bipy)_2Br_2^+$	–	–	0.070–0.126	–	–	0.032
cis-$Rh(bipy)_2I_2^+$	0.112	0.122	–	–	–	0.13
cis-$Rh(phen)_2Cl_2^+$	–	–	–	–	–	0.0096
cis-$Ir(phen)_2Cl_2^+$	–	–	0.020–0.036	–	–	0.014
cis-$Rh(phen)_2Br_2^+$	–	–	–	–	–	0.0016
cis-$Rh(phen)_2I_2^+$	–	–	–	–	–	0.052
trans-$Rh(py)_4Cl_2^+$	0.020	0.023	–	0.018	0.034	0.0094
trans-$Rh(py)_4Br_2^+$	–	~0.026	–	–	–	0.016
trans-$Ir(py)_4Cl_2^+$	–	–	0.01–0.018	–	–	0.0062
			Φ(py)			
trans-$Rh(py)_4Cl_2^+$	0.023	0.031	–	0.028	0.043	0.0092
trans-$Rh(py)_4Br_2^+$	–	~0.01	–	–	–	0.0085

[1]) Ref. [61]).

photoreduction to Co(II) is generally negligible compared to ligand substitution in the LF region. Some data are given in Table 11.[47]) Second, the ligand undergoing substitution generally agrees with the notions developed above concerning the one-electron population of the lowest excited state. Finally, note that the tetradentate cyclam ligand does yield a photosensitive complex compared to the *bis*-ethylene-

Table 11. Photosubstitution of Co(III)-amines upon 488nm photolysis at 25°[1])

Complex	pH	$\Phi NH_3 \times 10^4$	$\phi X^- \times 10^4$	ϕ Co(II) $\times 10^4$
$Co(NH_3)_6^{3+}$	2	3.2 ± 0.3	–	<0.035
$Co(NH_3)_5OH_2^{3+}$	2	1.8 ± 0.1	–	<0.050
$Co(NH_3)_5F^{2+}$	2	19.6 ± 1.5	5.5 ± 0.3	<0.38
$Co(NH_3)_5Cl^{2+}$	2	50.7 ± 1.3	17.1 ± 1.2	1.2
t-$Co(en)_2Cl_2^+$	2	None	10.7 ± 0.3	1.9
t-$Co(cyclam)Cl_2^+$	2	None	4.0 ± 0.1	<0.1
c-$Co(en)_2Cl_2^+$	2	~None	24	–

[1]) Ref.[47]).

diamine complex, in contrast to the Cr(III)systems. This further substantiates the notion of an inherently different mode of Cl^- substitution in the d^6 [Co(III), Rh(III)] and the Cr(III) amine systems.

A number of other low-spin d^6 complexes of the Group VIII metals have been subjected to photolysis and many have proven to be quite photosensitive. Among these are the O_h $Fe(CN)_6^{4-}$, $Ru(CN)_6^{4-}$, and $Pt(CL)_6^{2-}$ which are interpretable in terms of LF excited state photosubstitution.[2b] Being O_h no new stereochemical information is available here but some mechanistic insight may be possible. Next to $Co(CN)_6^{3-}$, $Fe(CN)_6^{4-}$ has received the most study of these O_h Group VIII low-spin d^6 systems. Interestingly, even nanosecond laser flash photolysis[64] did not reveal any discrete intermediates (either electronically excited states or coordinatively unsaturated intermediates) in photoaquation. Such a result is consistent with the mechanism proposed for $Co(CN)_6^{3-}$ photoaquation which was described as a dissociative interchange of CN^- for H_2O.[65] The interchange mechanism was proposed on the basis of an inability to directly form $Co(CN)_5X^{n-}$ ($X = N_3^-, I^-$) from photolysis of Co $(CN)_6^{3-}$ in aqueous solutions of X. By way of contrast, it appears that the loss of X in $Co(CN)_5X^{n-}$ complexes occurs by a dissociative type mechanism, but it is not known whether the reactivity of the photogenerated $Co(CN)_5^{2-}$ is the same as $Co(CN)_5^{2-}$ generated in the thermal reactions of $Co(CN)_5X^{3-}$. Extremely short excited state lifetimes for these d^6 systems in aqueous solutions at 298 K are inferred from the lack of luminescence and the flash photolysis results for $Fe(CN)_6^{4-}$. The short lifetimes preclude capture of an excited state to form a seven-coordinate intermediate as a mechanism for substitution. Thus, it is reasonable to invoke an essentially dissociative mechanism in most cases. One final point is worth making here. The photoaquation[2b] of $Pt(Cl)_6^{2-}$ reveals that the lability induced by population of $\sigma^*(e_g)$ is not compensenated by the depopulation of $\pi_1^*(t_{2g})$. Otherwise the complex would be inert to irradiation in the LF region.

Though LF states are generally responsible for ligand photosubstitution, CT states may also play some role in net photosubstitution pathways. A detailed study has been reported[66] for $Ru(NH_3)_5(pyridine)^{2+}$ and related complexes where a M → pyridine CT state is the lowest excited state in the molecule. Aside from the interesting fact that ligand substitution is accelerated upon photolysis into the M → pyridine CT transition, the $Ru(NH_3)_5(pyridine)^{2+}$ photosubstitution quantum efficiency was found to be dependent on the pH. Both *cis* and *trans* ammonia aquation is found, but unlike the pyridine photoaquation the ammonia photoaquation efficiency is not dependent on the pH. The most interesting mechanism to account for these results features a pH independent pathway via intersystem crossing from the CT state to a triplet LF state which gives Ru – N cleavage with small stereoselection since a nearly O_h system obtains. The increasing pyridine substitution yields with increasing pH are then rationalized by invoking protonation of a second CT excited state (probably triplet) which is also populated by intersystem crossing from the CT state achieved in absorption. Phenomenologically, the $Ru(NH_3)_5$ $pyridine^{2-}$ is unique because the substitution efficiency is apparently strongly influenced by intermolecular effects. A detailed interpretation of the results must await the results with more systems, but the possibilities here with respect to reactions of coordinated ligands with molecules of ions in the medium merit serious consideration.

3. Photosubstitution in d^6 Organometallic Complexes

A large number of d^6 organometallic complexes undergo ligand substitution, and this chemistry has been reviewed recently.[2a, c, d] Spectroscopically, the emerging fact is that the electronic spectra of these highly covalent complexes can be interpreted within the LF framework. Systems which have been given the LF assignment for the lowest energy absorption include $V(CO)_6^-$;[41] $M(CO)_6$ (M = Cr, Mo, W); $M(CO)_6^+$ (M = Mn, Re);[41] $M(CO)_5X$ (M = Mn, Re; X = Cl, Br, I; or M = Cr, Mo, W; X = *n*-electron donor);[67,68] *cis*-$M(CO)_4X_2$ (M = Cr, Mo, W; X = amine).[69] These and a few other systems have recently received some detailed study, and the essential results will be outlined here.

First, the O_h d^6 metal carbonyls undergo photosubstitution by a mechanism which is likely none other than simple dissociation of coordinated CO subsequent to the $t_{2g}^6 \rightarrow t_{2g}^5 e_g^1$ transition. The resulting $M(CO)_5$ species, reaction (11), is spectroscopically detectable by a number of

$$M(CO)_6 \xrightarrow{h\nu} M(CO)_5 + CO \qquad (11)$$

Table 12. Photosubstitution in d^6 $M(CO)_5X$ complexes [1)]

M	X	436 nm		366 nm		313 nm	
		ϕ_X	ϕ_{CO}	ϕ_X	ϕ_{CO}	ϕ_X	ϕ_{CO}
Mo	Piperidine	–	0.04_8	–	0.13	–	–
Mo	NH_3	0.58	–	0.74	–	–	–
Mo	*n*-$PrNH_2$	–	0.05_7	–	0.24	–	–
W	NH_3	0.56	–	0.49	–	–	–
W	*n*-$PrNH_2$	0.73	0.01_6	0.60	0.05_7	–	–
W	Piperidine	0.49	0.00_6	0.45	0.03_4	–	–
Re	Cl	–	–	~ 0.00	0.20	~ 0.00	0.76
Re	Br	–	–	~ 0.00	0.34	~ 0.00	0.58
Re	I	–	–	~ 0.00	0.10	~ 0.00	0.67

1) Refs.[68b, 70].

techniques for M = Cr, Mo, W.[2a] More recently, there has been a report[42] of generation of unstable intermediates via photolysis of $M(CO)_6^-$ (M = V, Nb, Ta) in low temperature glassy media or in KBr pressed discs. No reports concerning the photochemistry of the $Mn(CO)_6^+$ or $Re(CO)_6^+$ have appeared. The dissociative loss of CO from the neutral d^6 hexacarbonyls is believed to occur with unit quantum efficiency[2a] but all that can be said is that the observed labilization is consistent with the change in one-electron configuration that occurs upon population of the low lying LF states.

Detailed study of the photosubstitution in substituted d^6 metal carbonyls has been reported for $M(CO)_5(\text{amine})$[70] (M = Cr, Mo, W) and $M(CO)_5X$ (M = Re, X = Cl, Br, I).[68b] Some quantum yield data for these reactions are shown in Table 12 and some electronic spectra are given in Fig. 2. Despite the remarkable similarity in the

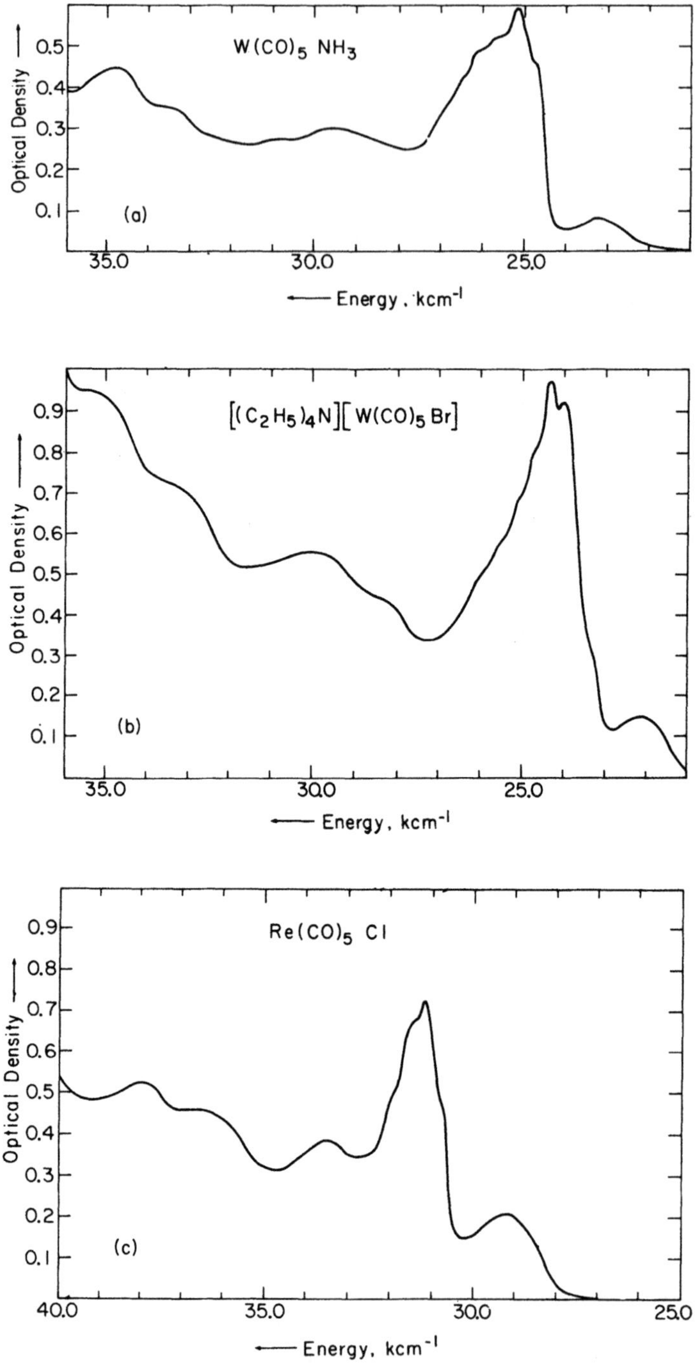

Fig. 2. Comparison of the electronic spectra of indicated complexes at 25 K in EtOH. The first absorption band has a molar absorptivity of 600, 500, and 750 $1\,mol^{-1}\,cm^{-1}$ in solution at 298 K for $W(CO)_5NH_3$, $[W(CO)_5Br]^-$, and $Re(CO)_5Cl$, respectively

electronic spectra of the $M(CO)_5X$ complexes, the photochemistry of the $M(CO)_5$(amine) involves mainly photosubstitution of amine while $Re(CO)_5X$ photosubstitution involves CO. However, the following similarity does remain: photolysis into the lowest energy portion of the first absorption band leads to substantially lower CO substitution quantum yields than upon photolysis into the high energy side of the first major band maximum. These facts are seemingly rationalized by the same reasoning used in the Group VIII d^6 complexes. The lowest absorption system in these C_{4v} systems likely involve a transition resulting in the population of the $d_{z^2}(a_1)\sigma_z^*$ orbital at the expense of the d_{xz}, d_{yz} $(e)\pi$-d orbitals. Thus, the OC-M-X axis is substantially labilized and in the case of X = amine; M = Mo, W this axis of labilization does obtain and the small CO substitution yield involves the substitution of the axial-CO. In the $Re(CO)_5X$ case the axial-CO is likely lost exclusively due to the compensating π-donor interaction of Cl, Br, and I which can be more fully exploited by depopulation of the π-d orbital. The more favorable π-donor situation in the excited state diminishes the effect on Re-X lability by the population of $d_{z^2}(a_1)\sigma_z^*$. Consequently, the Re-CO bond is substantially more labilized especially since depopulation of the π-d level *diminishes* π-bonding for the Re-CO interaction. In both $Re(CO)_5X$ and $M(CO)_5$(amine) higher energy photolysis leads to population of LF states which feature population of the $d_{x^2-y^2}(b_1)\sigma_{xy}^*$ orbital which labilizes the equatorial CO's and leads to larger CO substitution quantum yields. In all of these C_{4v} complexes the ligand photosubstitution most likely occurs by strictly a dissociative mechanism to yield coordinatively unsaturated intermediates. For the $Re(CO)_5X$, photolysis in the absence of added nucleophiles yields the dimeric species $[Re(CO)_4X]_2$, reaction (12), which likely form via coupling of two coordinatively unsaturated $Re(CO)_4X$ intermediates.[68b)]

$$Re(CO)_5X \xrightarrow{h\nu} (OC)_4Re(\mu\text{-}X)_2Re(CO)_4 + 2\,CO \qquad (12)$$

The comparison of the photosubstitution of $W(CO)_4$(1,10-phenanthroline) and *cis*-$W(CO)_4$(pyridine)$_2$ is also very interesting.[71)] The electronic spectra of these two species are very different. The lowest excited state in $W(CO)_4$(1,10-phenanthroline) is a W → 1,10-phenanthroline CT absorption while the lowest excited state in *cis*-$W(CO)_4$(pyridine)$_2$ is LF and is interpreted according to Scheme 7. The photosubstitution quantum yields are summarized in Table 13.[71)] The chemistry of the $W(CO)_4$(1,10-phenanthroline) involves CO substitution, but apparently the photochemistry does not originate in the lowest excited state. The quantum yield increases by two orders of magnitude upon increasing the excitation energy and the increase in quantum yields correlates with the position of the LF region. The loss of the 1,10-phenanthroline ligand is unlikely as it is a 4-electron donor and is seemingly incapable of being a 2-electron donor. However, photolysis of the *cis*-$W(CO)_4$(pyridine)$_2$ in the LF region leads very efficiently to loss of pyridine. Yields for CO release in the same region for the *bis*-pyridine complex are of the order of 10^{-3}. In the *bis*-pyridine complex the OC-W-N axes are labilized in the lowest excited state, and, as in

Table 13. Photosubstitution in $W(CO)_4$(5-4-1,10-phenanthroline) and *cis*-$W(CO)_4(pyridine)_2$

Y	Rxn.	X(conc)	ϕ_{436}	ϕ_{405}	ϕ_{366}	ϕ_{313}
H	A	CH_3CN (neat)	1.6×10^{-4}	1.2×10^{-3}	9.2×10^{-3}	2.2×10^{-2}
CH_3	A	CH_3CN (neat)	1.5×10^{-4}	–	–	2.0×10^{-2}
Br	A	CH_3CN (neat)	1.2×10^{-4}	–	–	2.4×10^{-2}
–	B	–	2.3×10^{-1}	2.7×10^{-1}	2.3×10^{-1}	–

Reactions: A. $W(CO)_4$(5-Y-1,10-phenanthroline) $\xrightarrow[X]{h\nu}$ $XW(CO)_3$(5-Y-1,10-phenanthroline) + CO

: B. *cis*-$W(CO)_4(pyridine)_2$ $\xrightarrow[\text{1,10-phenanthroline}]{h\nu}$ $W(CO)_4$(1,10-phenanthroline) + 2 pyridine.

Ref.[71].

$W(CO)_5$(pyridine) where the OC-W-N axis is also labilized in the lowest excited state, efficient pyridine photosubstitution obtains.

The photochemistry of the isoelectronic series (η^6-benzene)$Cr(CO)_3$, (η^5-cyclopentadienyl)$MN(CO)_3$, and (η^4-cyclobutadienyl)$Fe(CO)_3$ all involve CO photosubstitution as their principal primary photoprocess.[2a)] None of these are thought to undergo substitution of the 6-electron donor π-system with high quantum efficiency as previously suspected. The similarity in reactivity shows the value of isoelectronic relationships in these complexes, but no detailed treatment of the electronic structure of d^6 (arene)$M(CO)_3$ complexes has appeared.

C. Photosubstitution in d^8 Complexes

By comparison to the d^3 and low-spin d^6 systems, d^8 complexes have received considerably less study by photochemists. The electronic structure of both 4- and 5-coordinate d^8 complexes has been an area of intense activity,[72)] however, and the results are of use in the interpretation of the photochemistry. In the 4- and 5-coordinate d^8 complexes the ordering of the occupied d-orbitals has been a point of controversy in several cases, but the nature of lowest unoccupied d-orbital is unequivocal. For example, in $PtCl_4^{2-}$ and related complexes the d_{z^2}, d_{xz}, d_{yz}, and d_{xy} orbitals are all occupied and $d_{x^2-y^2}$ is unoccupied and is strongly σ-antibonding. Therefore, all LF transitions should lead to a destabilization of the Pt-Cl interaction with respect to

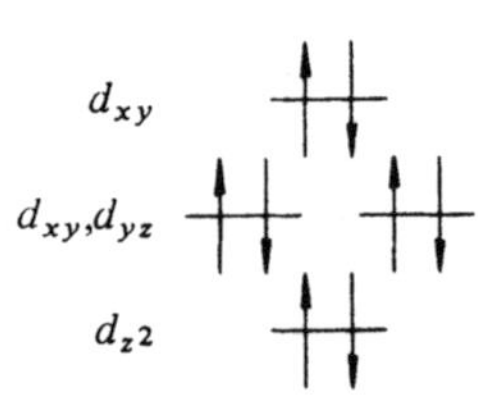

$PtCl_4^{2-}$

Scheme 9

the σ-bonding.[5] The ordering of the occupied *d*-orbitals now seems well established for $PtCl_4^{2-}$ and is shown in Scheme 9. The photosubstitution of $PtCl_4^{2-}$ has been studied in some detail[73] and reaction (13) occurs in aqueous solution upon photo-

$$PtCl_4^{2-} \xrightarrow{h\nu} PtCl_3(OH_2)^- + Cl^- \tag{13}$$

lysis in the LF region. The reaction quantum yields are included in Table 14 and are seen to be essentially independent of the LF state achieved, but upon 254 nm excitation, which presumably results in CT excitation, the quantum yield increases substantially.

Table 14. Photosubstitution in d^8 complexes

Complex	Photoproduct	$\phi(\lambda, nm)$	Ref.
$PtCl_4^{2-}$	$PtCl_3(OH_2)^-$	0.17 (472); 0.19 (404); 0.20 (313); 0.90 (254)	73)
$PtCl_3(C_2H_4)^-$	$PtCl_3(OH_2)^-$	$<10^{-4}$ (420); $<10^{-4}$ (410); 0.008 (385); 0.066 (345); 0.10 (305)	74)
	cis-$PtCl_2(OH_2)(C_2H_4)$	~ 0.15 (340); ~0.14 (305)	
$Fe(CO)_5$	$Fe(CO)_4 + CO$	–	76)
$Ru(CO)_5$	$Ru_2(CO)_9$	–	77)
$Os(CO)_5$	$Os_2(CO)_9$	–	77)
$(\eta^5\text{-}C_5H_5)Rh(CO)_2$	$(\eta^5\text{-}C_5H_5)_2Rh_2(CO)_3$	–	78)
$Mn(CO)_4NO$	$Mn(CO)_3NO + CO$	–	79)
$RCo(CO)_4$	$RCo(CO)_3PF_3$	–	80)

The photochemistry of $PtCl_3(C_2H_4)^-$ has also been studied in some detail and some interesting observations have been made involving C_2H_4 photosubstitution.[74] The C_2H_4 substitution is remarkable because this is not a reaction which takes place thermally. Additionally, the substitution of C_2H_4 occurs with a strong wavelength dependence, Table 14. The increasing photosubstitution yield with increasing excitation energy was interpreted by Wrighton, Gray and Hammond[5] using an orbital scheme as shown in Scheme 10 below. Since all LF transitions result in the population of $d_{x^2-y^2}$ the only rationale within the LF framework for variation in photo-

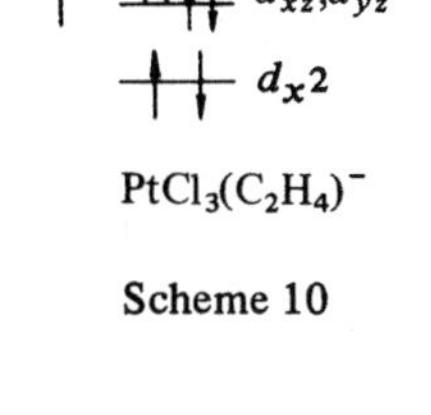

$PtCl_3(C_2H_4)^-$

Scheme 10

substitution efficiency is to ascribe different labilities depending on the orbital *depopulated.* In particular, depopulation of d_{xz} or d_{yz} (depending on the axis choice) diminishes the π-bonding between Pt and C_2H_4 at the same time that $d_{x^2-y^2}$ population reduces the σ-bonding. Such a transition was reasoned to fall higher in energy than the $d_{xy} \rightarrow d_{x^2-y^2}$ transition and hence C_2H_4 substitution would more likely occur at the higher excitation energies. Subsequently, Rösch, Messmer, and Johnson carried out an SCF-Xα scattered wave calculation on $PtCl_3(C_2H_4)^-$ and concluded that the lowest absorptions are not LF, but in fact, are more appropriately identified as Cl → Pt CT absorptions.[10] A higher energy Pt → C_2H_4 absorption involving population of an orbital antibonding with respect to Pt-C_2H_4 was calculated to fall near the observed onset of C_2H_4 photosubstitution.

One mechanistic study[75] worth describing here concerns the photoreactivity of $[Pt(diethylenetriamine)Br]^+$. Photolysis in the presence of NO_2 accelerates the substitution of Br^- to yield $[Pt(diethylenetriamine)NO_2]^+$. The reaction was shown to proceed *via* $[Pt(diethylenetriamine)OH_2]^{2+}$ which is rapidly anated by either Br^- or NO_2^-. The essential evidence rests in the fact that photolysis in basic solution yields only $[Pt(diethylenetriamine)OH]^+$ even in the presence of NO_2^-. This result prompts the postulate that a dissociative interchange mechanism obtains as proposed for $Co(CN)_6^{3-}$.[65]

Most other work with d^8 complexes concerns 5-coordinate organometallic complexes. Some examples are given in Table 14.[73, 74, 76–80] The electronic spectrum of $Fe(CO)_5$ has been measured and the lowest absorption has been identified as LF $d_{xy}, d_{x^2-y^2} \rightarrow d_{z^2}$ according to the one-electron level scheme in Scheme 11.[81] Such a transition should yield labilization principally along the z-axis due to the

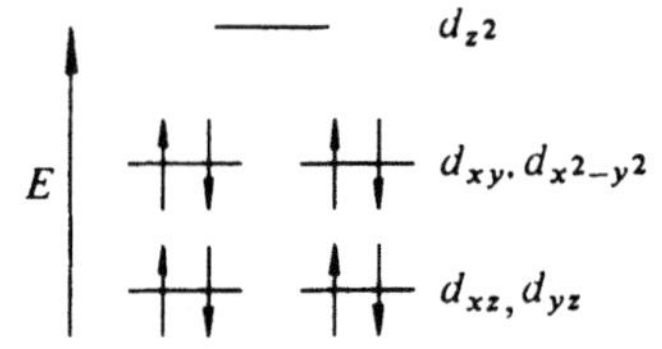

$Fe(CO)_5$

Scheme 11

directed σ-antibonding character of d_{z^2}. Electronic excitation of $Fe(CO)_5$ does, of course, induce CO substitution and recently the $Fe(CO)_4$ product from CO photodissociation has been observed and identified spectroscopically.[76] Other examples of CO substitution in d^8 metal carbonyls are given in Table 14. These examples seem rather straightforward except for $Mn(CO)_4NO$ where an entering group concentration dependence has been observed for the CO substitution quantum yield.[79] In this interesting case the quantum yield for $Mn(CO)_3(PPh_3)NO$ formation increases with increasing PPh_3 concentration. The isomerization of NO from linear to bent (three-to one-electron donor) could be invoked as a primary photoprocess, with substitution proceeding according to reactions (14)–(16).

$$(OC)_4Mn{-}N{=}O \xrightarrow{h\nu} [(OC)_4Mn{-}N{=}\dot{O}] \quad (14)$$

"coordinatively unsaturated"

$$PPh_3 + [(OC)_4Mn{-}N{=}O] \rightarrow [PPh_3(OC)_4Mn{-}N{=}O] \quad (15)$$

$$[(PPh_3)(OC)_4Mn{-}N{=}O] \rightarrow PPh_3(CO)_3Mn{-}N{=}O + CO \quad (16)$$

D. Applications of Photosubstitution

Light induced ligand substitutions have several chemical and physical applications. Chemically, photosubstitution reactions have provided a valuable synthetic route to substituted derivatives of thermally inert complexes. This fact is particularly true in the metal carbonyl area.[2a)] Moreover, the photosubstituted products can be generated at low enough temperatures to be isolable whereas if generated thermally at elevated temperatures the products may be unstable to the reaction conditions. One example here is photosubstitution of $Fe(CO)_5$ by pyridine to yield $Fe(CO)_4^-$ (pyridine)[82)] at 25 °C whereas thermal reaction at higher temperatures yields disproportionation of the $Fe(CO)_5$.[83)] The stereochemistry of photosubstitution is also potentially different from the thermal reactivity and a typical example here is the loss of SO_3^{2-} from *trans*-$Co(CN)_4(SO_3)(OH_2)^{3-}$ which is a reaction which does not occur thermally.[54)] In a similar way C_2H_4 in $PtCl_3(C_2H_4)^-$ is substantially inert thermally, but can be substituted photochemically.[74)] Photoexcitation of (arene)$Cr(CO)_3$ leads to dissociative loss of CO while thermally the arene group is labilized.[84)]

Photodissociation of coordinated ligands has also lead to the synthesis of new complexes by oxidative addition to the coordinatively unsaturated intermediates as in reactions (17)–(19).[85–87)] These types of reactions have been invoked in transition metal complex photoassisted and photocatalyzed reactions.

$$Fe(CO)_5 \xrightarrow[R_3Si\text{-}H]{h\nu} cis\text{-}HFe(CO)_4SiR_3 + CO \quad (17)^{85)}$$

$$Os(CO)_3(PPh_3)_2 \xrightarrow[H_2]{h\nu} H_2Os(CO)_2(PPh_3)_2 + CO \quad (18)^{86)}$$

$$(\eta^5\text{-}C_5H_5)Rh(CO)_2 \xrightarrow[R_3Si\text{-}H]{h\nu} (\eta^5\text{-}C_5H_5)Rh(CO)HSiR_3 + CO \quad (19)^{87)}$$

Simple photosubstitution processes in metal complexes may also lead to useful photochromic systems. The general concept may be a reduction in symmetry and a lowering of average LF strength, as in reaction (20), leading to changes in the position of the LF bands. Additionally, one can take advantage of substantial changes in the nature of CT transitions depending on the new complexes as in reaction (21).[88, 89)]

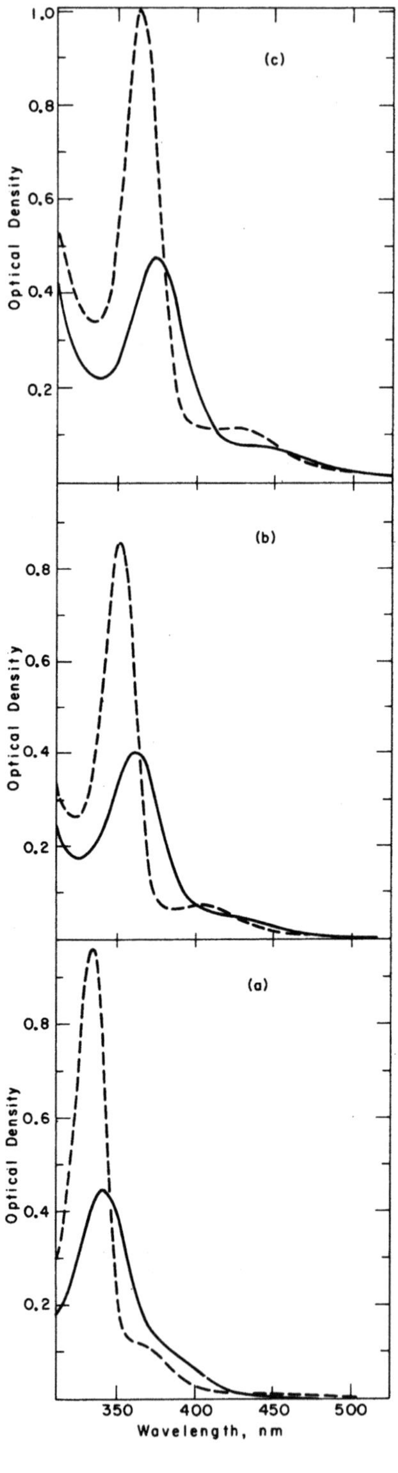

Fig. 3. Electronic absorption spectra of (a) $Mn_2(CO)_{10}$; (b) $Mn_2(CO)_9PPh_3$, and (c) $Mn_2(CO)_8$-$(PPh_3)_2$ in EPA at 298 (——) and 77 K (----). The sharp, intense peak ($\epsilon \approx 2x10^4$) is the $\sigma_b \rightarrow \sigma^*$ absorption and the low energy absorption shoulder is the π-$d \rightarrow \sigma^*$ (from Ref.[98])

$$\underset{\text{colorless}}{M(CO)_6} \underset{\Delta}{\overset{h\nu}{\rightleftarrows}} \underset{\text{yellow}}{M(CO)_5} + CO \qquad (20)^{88)}$$

$$\underset{\text{colorless}}{W(CO)_6} \underset{?}{\overset{h\nu}{\underset{\text{TCNE}}{\longrightarrow}}\atop{\longleftarrow}} \underset{\text{blue}}{W(CO)_5TCNE} + CO \qquad (21)^{89)}$$

In an area of considerably less understanding, photosubstitution chemistry may play a key role in elucidating certain steps in heterogeneous catalysis. There now exist a number of reports where photodesorption and photoadsorption to catalytic surfaces occurs.[90, 91] These may be viewed as complicated photosubstitution reactions and there are some reports that these reactions can measurably influence the rate of the catalytic transformation of substrates.[90, 91]

II. Photodeclusterification of Metal-Metal Bonded Complexes

The metal-metal bond is a very important structural unit in chemistry and has been well characterized for a large number of systems by X-ray crystallographic structure determinations.[92] Multiple metal-metal bonds exist and metal atoms can be bonded to more than one other metal. Further, the metal-metal bond need not involve the same two metals. The metal-metal bond tends to be found for low valent, heavy, early transition elements. Metal carbonyls are remarkable in that metal-metal bonding is very important for nearly every metal which forms complexes with CO. The strength of metal-metal bonds is a matter of some debate in the literature with bond energies ranging from ~11 kcal/mole for the Co–Co single bond in $Co_2(CO)_8$[93] to ~375 kcal/mole[94] for the quadruple Re-Re bond in $Re_2Cl_8^{2-}$. Despite what the finer details of electronic structure may be, all workers in this area seem to agree that there exist low lying excited electronic configurations in metal-metal bonded complexes which feature a change in the metal-metal interaction compared to the ground electronic configuration.[95] This fact confronts us with a very exciting prospect: photolysis may result in clean metal-metal bond cleavage. Such a reaction has

Table 15. Photodeclusterification of metal carbonyls

Starting Complex(es)	Medium	Photoproduct(s)	Ref.
$Mn_2(CO)_{10}$	CCl_4	$Mn(CO)_5Cl$	98, 99)
$Re_2(CO)_{10}$	CCl_4	$Re(CO)_5Cl$	98, 100)
$Mn_2(CO)_{10}$	Isooctane/I_2	$Mn(CO)_5I$	98)
$Re_2(CO)_{10}$	Isooctane/I_2	$Re(CO)_5I$	98)
$Mn_2(CO)_{10} + Re_2(CO)_{10}$	Isooctane	$MnRe(CO)_{10}$	98, 101)
$Mn_2(CO)_9PPh_3$	Isooctane	$Mn_2(CO)_{10} + Mn_2(CO)_8(PPh_3)_2$	98)
$[(\eta^5\text{-}C_5H_5)Mo(CO)_3]_2$	CCl_4	$(\eta^5C_5H_5)Mo(CO)_3Cl$	102)
$[(\eta^5\text{-}C_5H_5)W(CO)_3]_2$	CCl_4	$(\eta^5\text{-}C_5H_5)W(CO)_3Cl$	102)
$[(\eta^5\text{-}C_5H_5)W(CO)_3]_2$ + $Mn_2(CO)_{10}$	Isooctane	$(\eta^5\text{-}C_5H_5)W(CO)_3\text{-}Mn(CO)_5$	102)
$[(\eta^5\text{-}C_5H_5)Mo(CO)_3]_2$ + $Mn_2(CO)_{10}$	Isooctane	$(\eta^5\text{-}C_5H_5)Mo(CO)_3\text{-}Mn(CO)_5$	102)
$(\eta^5\text{-}C_5H_5)Mo(CO)_3\text{-}Mn(CO)_5$	Isooctane	$Mn_2(CO)_{10}+[(\eta^5\text{-}C_5H_5)Mo(CO)_3]_2$	103)
$Co_2(CO)_8$ + 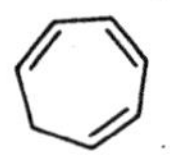		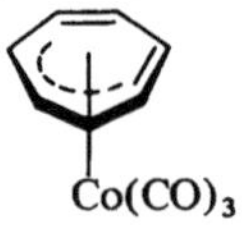	104)
$Os_3(CO)_{12}$	1,5-Cyclo-octadiene	(1,3-Cyclooctadiene)$Os(CO)_3$	105)
	$(CH_3)_3SiH$	$(CH_3)_3SiOs(CO)_4H$	106)
$Ru_3(CO)_{12}$	Benzene/CO	$Ru(CO)_5$	107)
+	$(CH_3)_5SiH$	$[(CH_3)_3Si]_2Ru(CO)_4$	106)
$Mn_2(CO)_{10}$ + $[(\eta^5\text{-}C_5H_5)Fe(CO)_2]_2$	Benzene	$(\eta^5\text{-}C_5H_5)Fe(CO)_2\text{-}Mn(CO)_5$	108)

some precedence thermally,[96] but in any case the reaction is not well studied. Cleavage of the metal-metal bond by photolysis may be a successful route to extremely reactive intermediates of a radical nature and may have some analogy to photoinduced homolytic cleavage[97] in peroxides, disulfides, halogens, etc.

A. Photodecludsterification of Metal Carbonyls

Quite a number of dinuclear, and three trinuclear, metal carbonyls have received some study with respect to photoinduced metal-metal bond cleavage. Some typical systems are given in Table 15.[98–108] First, consider $Mn_2(CO)_{10}$ which is perhaps

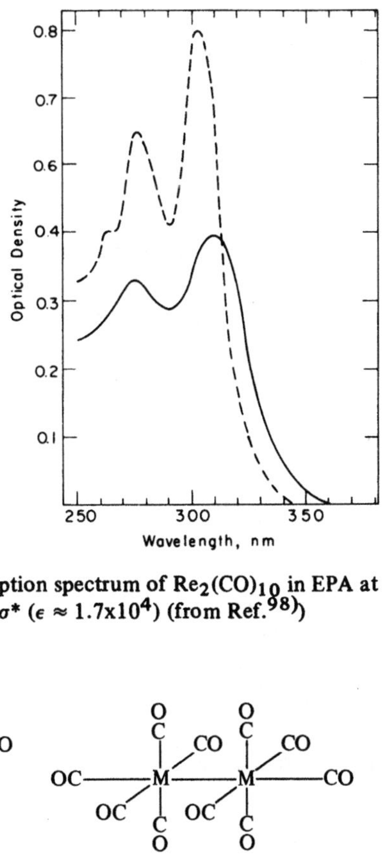

Fig. 4. Electronic absorption spectrum of $Re_2(CO)_{10}$ in EPA at 298 (——) and 77 K (---). Band at ~308 nm is the $\sigma_b \rightarrow \sigma^*$ ($\epsilon \approx 1.7 \times 10^4$) (from Ref.[98])

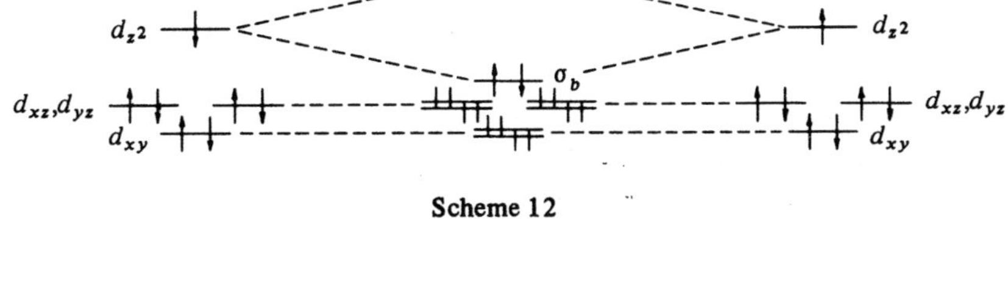

Scheme 12

the most well understood of these several systems. The electronic picture has been determined to be as in Scheme 12 with experimental evidence coming mainly from polarized electronic absorption spectra.[109] The spectra are shown in Fig. 3 and the low energy shoulder is ascribed to a π-$d \rightarrow \sigma^*$ transition while the intense near uv absorption maximum is associated with the $\sigma_b \rightarrow \sigma^*$ transition according to Scheme 12.[109] The spectrum of $Re_2(CO)_{10}$, Fig. 4, is interpreted in a similar way. In the one-electron scheme, then, it is clear that electronic excitation will destabilize the M–M interaction in these d^7-d^7 systems.

Recently, some detailed and quantitative studies of the photochemistry of these d^7-d^7 metal carbonyls have been reported.[98] The chemistry obtained is consistent with efficient photoinduced, homolytic metal-metal bond cleavage. In fact, several lines of chemical evidence support such a pathway for the relaxation of the low lying electronically excited states of these complexes. First, photolysis of a mixture of $Mn_2(CO)_{10}$ and $Re_2(CO)_{10}$ leads to the heterodinuclear species, reaction (22). This product seemingly results from cross-coupling of a $Mn(CO)_5$ and a $Re(CO)_5$

$$Mn_2(CO)_{10} + Re_2(CO)_{10} \xrightarrow[\text{alkane}]{h\nu} MnRe(CO)_{10} \qquad (22)$$

radical since the formation of $MnRe(CO)_{10}$ *requires the photolysis of both $Mn_2(CO)_{10}$ and $Re_2(CO)_{10}$*. Naturally, $MnRe(CO)_{10}$ is only an initial photoproduct as it also absorbs light and undergoes Mn-Re cleavage to regenerate the homodinuclear species. The important point is that pure $MnRe(CO)_{10}$ having the electronic spectrum given in Fig. 5 initially yields a 1 : 1 ratio of $Mn_2(CO)_{10}$ and $Re_2(CO)_{10}$ upon photolysis, reaction (23). This initial ratio of the products strongly supports the photogeneration of metal radicals which couple in an essentially statistical fashion.

$$MnRe(CO)_{10} \xrightarrow[\text{alkane}]{h\nu} \underset{1}{Mn_2(CO)_{10}} \; \underset{:}{+} \; \underset{1}{Re_2(CO)_{10}} \qquad (23)$$

A second line of chemical evidence in support of homolytic cleavage of Mn-Mn, Re-Re, and Mn-Re bonded metal carbonyls comes from examination of the photoproducts and the efficiency of their formation in the presence of halogen donors. Some convincing data are summarized in Table 16.[98, 110] Generally, it appears that the reactions follow the stoichiometry indicated in reaction (24) and in many cases the quantum yields are very high. For the halogen donors Ph_3CCl and $PhCH_2Cl$ the

$$(L)_{5-m}(OC)_m M{-}M'(CO)_n(L)_{5-n} \xrightarrow[\substack{\text{halogen}\\ \text{donor}\\ \text{X=halogen}}]{h\nu} XM(CO)_m(L)_{5-m} + XM'(CO)_n(L)_{5-n} \qquad (24)$$

organic radical or its coupling product has been observed. Importantly, with CH_3I as the halogen donor for $Re_2(CO)_{10}$ only $Re(CO)_5I$ was found as the metal containing product. The reaction stoichiometry and quantum efficiency in (24) is consistent with

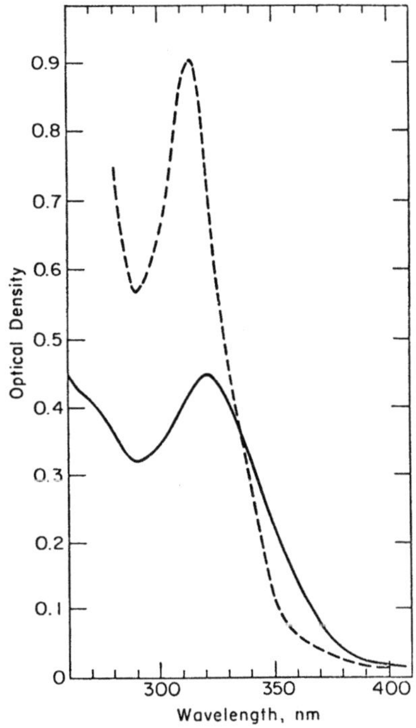

Fig. 5. Electronic absorption spectra of $ReMn(CO)_{10}$ in EPA at 298 (——) and 77 K (---). Band at ≈ 315 nm is the $\sigma_b \rightarrow \sigma^*$ ($\epsilon \approx 1.5 \times 10^4$) (from Ref.[98])

photoinduced homolytic cleavage to yield d^7 $M(CO)_n L_{5-n}$ (M = Mn, Re) type intermediates. These d^7 intermediates would be expected to have some reactivity properties in common with the d^7 $Co(CN)_5^{3-}$ which does react readily with halogen donors by a free radical-like mechanism.[111]

The only claim of direct evidence for the intermediates proposed above come from ESR measurements subsequent to photolysis of $Mn_2(CO)_{10}$ in tetrahydrofuran at room temperature.[112] A long-lived, ESR detectable, radical was found and proposed to be $Mn(CO)_5 \cdot THF$. The intermediate disappears upon addition of I_2 and the formation of $Mn(CO)_5I$ is observed. These data seem to be wholly consistent with the photochemistry outlined above, but the interpretation of the ESR signal as that due to an $Mn(CO)_5$ moiety seems untenable because it is too long-lived. The $Re(CO)_5$ species proposed as an intermediate in the photolysis of $Re_2(CO)_{10}$ has recently been synthesized by atom/ligand co-condensation synthesis and infrared data in the matrix at low temperature support a square-pyramidal structure.[113] An ESR signal was also observed from a species thought to be $Mn(CO)_5$ formed by subliming $Mn_2(CO)_{10}$ on to a cold tip.[114] The ESR detectable species is now believed to be $\cdot OOMn(CO)_5$.[115]

Photosubstitution also occurs in $Mn_2(CO)_{10}$[116] but the reaction likely involves first the photogeneration of the mononuclear $Mn(CO)_5$ species.[98] The evidence for this statement comes from the fact that $Mn_2(CO)_8(PPh_3)_2$ is a *primary photoproduct* upon photolysis of $Mn_2(CO)_{10}$ in the presence of PPh_3. The reaction could proceed as indicated in Scheme 13. The key here is that substitution occurs thermally and the final products are generated by coupling of the paramagnetic fragments.

Table 16. Photolysis of Mn and Re dinuclear carbonyls in the presence of halogen donors

Starting complex	Halogen donor	Product	Irrdn λ, nm	ϕ_{dis}[1]	$\phi_{formn.}$[2]
$Mn_2(CO)_{10}$[3]	CCl_4	$Mn(CO)_5Cl$	313	0.48	1.02
			366	0.41	0.72
	I_2	$Mn(CO)_5I$	366	0.44	0.79
	Ph_3CCl	$Mn(CO)_5Cl$ + $Ph_3C\cdot$	Near UV	–	–
	$PhCH_2Cl$	$Mn(CO)_5Cl$ + $(PhCH_2)_2$	Near UV	–	–
$MnRe(CO)_{10}$[3]	CCl_4	$Re(CO)_5Cl$ + $Mn(CO)_5Cl$	366	0.42	0.43 0.46
	I_2	$Re(CO)_5I$ + $Mn(CO)_5I$	366	0.45	0.38 0.37
$Re_2(CO)_{10}$[3]	CCl_4	$Re(CO)_5Cl$	313	0.60	1.20
	I_2	$Re(CO)_5I$	313	0.64	1.14
	Ph_3CCl	$Re(CO)_5Cl$ + $Ph_3C\cdot$	Near UV	–	–
	$PhCH_2Cl$	$Re(CO)_5Cl$ + $(PhCH_2)_2$	Near UV	–	–
	CH_3I	$Re(CO)_5I$	Near UV	–	–
$Mn_2(CO)_9PPh_3$[3]	CCl_4	$Mn(CO)_5Cl$ $Mn(CO)_4PPh_3Cl$	366	0.45	0.36 0.40
$Mn_2(CO)_8$-(1,10-phenanthroline)[4]	Chlorinated Solvent	$Mn(CO)_5Cl$ + $ClMn(CO)_3$-(1,10-phenanthroline)	UV	–	–

[1] Quantum yield for disappearance of starting complex, ± 10%.
[2] Quantum yield for formation of product, ± 10%.
[3] Ref.[98].
[4] Ref.[110].

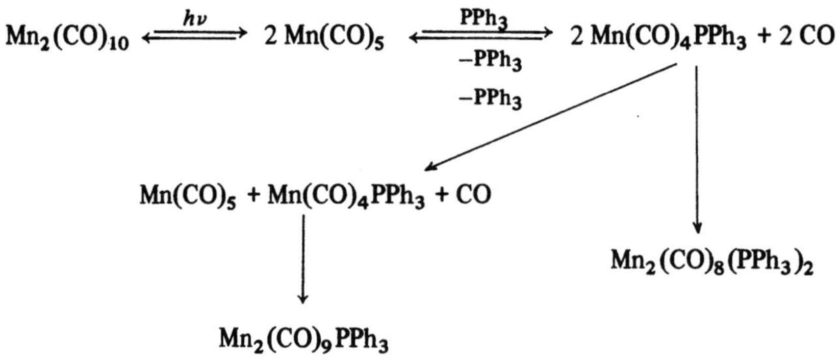

Scheme 13

Another set of metal-metal bonded complexes that have received some detailed study[102,117–119] are the d^5-d^5 $[(\eta^5\text{-}C_5H_5)M(CO)_3]_2$ (M = Mo, W) complexes. Early work indicated that photolysis in the presence of PPh_3 (and related ligands) could yield simple ligand substitution products, reaction (25).[117] This is to be contrasted with the observation of net disproportionation as indicated in reaction (26).[118] And

$$(\eta^5\text{-}C_5H_5)_2Mo_2(CO)_6 \xrightarrow[PPh_3]{h\nu} (\eta^5\text{-}C_5H_5)_2Mo_2(CO)_5PPh_3 \quad (25)$$

$$[(\eta^5\text{-}C_5H_5)Mo(CO)_3]_2 \xrightarrow[PPh_3]{h\nu} [(\eta^5\text{-}C_5H_5)Mo(CO)_2(PPh_3)_2^+] + [(\eta^5\text{-}C_5H_5)Mo(CO)_3^-] \quad (26)$$

recently there has been a claim[119] that heterolytic cleavage of the Mo–Mo bond can result by an efficient decay path of the excited $[(\eta^5\text{-}C_5H_5)Mo(CO)_3]_2$ complex to generate the metal carbonyl anion, $(\eta^5\text{-}C_5H_5)Mo(CO)_3^-$. To all of this qualitative photochemistry we must also add the fact that photolysis of a mixture of $Mn_2(CO)_{10}$ and $[(\eta^5\text{-}C_5H_5)M(CO)_3]_2$ yields the heterodinuclear $(OC)_5Mn\text{–}M(CO)_3(\eta^5\text{-}C_5H_5)$ complex, reaction (27).[102] This photoreaction occurs only when *both* homodinuclear starting materials absorb the incident irradiation. The chemical efficiency for reaction (27) is quite good, but the product is susceptible to back photoreaction (*vide infra*).

$$M_2(CO)_{10} + [(\eta^5\text{-}C_5H_5)M'(CO)_3] \xrightarrow[\text{isooctane}]{h\nu} (OC)_5M\text{–}M'(CO)_3(\eta^5\text{-}C_5H_5) \quad (27)$$

M = Mn, Re
M′= Mo, W

Some quantitative photochemical data for the $[(\eta^5\text{-}C_5H_5)M(CO)_3]_2$ species in the presence of halogen donors are given in Table 17.[102] The data in CCl_4 are particularly useful in that the quantum efficiency for disappearance of the M–M bonded species is quite high and the yield of $(\eta^5\text{-}C_5H_5)M(CO)_3Cl$ is essentially quantitative. Thus, it seems most reasonable that the primary photoprocess again involves homolytic cleavage of the M–M bond as indicated in reaction (28). The paramagnetic, mo-

$$[(\eta^5\text{-}C_5H_5)M(CO)_3]_2 \xrightarrow{h\nu} 2\,(\eta^5\text{-}C_5H_5)M(CO)_3 \quad (28)$$

nonuclear species thus generated can recouple with each other, couple with other metal radicals such as $Mn(CO)_5$, react with halogen donors to yield mononuclear metal carbonyl chlorides, and presumably can undergo substitution by nucleophiles such as PPh_3. Heterolytic cleavage, if it occurs at all, seemingly can only account for a minor fraction of the *primary* excited state decay process.

The photochemistry of the $[(\eta^5\text{-}C_5H_5)M(CO)_3]_2$ species is very similar to that associated with the d^7-d^7 Mn and Re metal-metal bonded complexes. The electronic spectra of the d^5-d^5 molecules are also remarkably similar to the d^7-d^7 systems: a low energy, low intensity band in the visible and an intense, fairly sharp near UV band. The spectra for $[(\eta^5\text{-}C_5H_5)M(CO)_3]_2$ (M = Mo, W) are shown in Fig. 6. The

Table 17. Irradiation of $[(\eta^5\text{-}C_5H_5)M(CO)_3]_2$ in the presence of halogen donors[1])

M	Halogen donor	Product(s)	Irrdn, λ nm	ϕ[2])
Mo	CCl_4	$(\eta^5\text{-}C_5H_5)Mo(CO)_3Cl$	366	0.45
			405	0.42
			550	0.35
	Ph_3CCl	$(\eta^5\text{-}C_5H_5)Mo(CO)_3Cl$ + $Ph_3C\cdot$	366 or 550	–
	$PhCH_2Cl$	$(\eta^5\text{-}C_5H_5)Mo(CO)_3Cl$ + $(PhCH_2)_2$	Near UV	–
W	CCl_4	$(\eta^5\text{-}C_5H_5)W(CO)_3Cl$	366	0.21
			550	0.12
	Ph_3CCl	$(\eta^5\text{-}C_5H_5)W(CO)_3Cl$ + $Ph_3C\cdot$	366 or 550	–
	$PhCH_2Cl$	$(\eta^5\text{-}C_5H_5)W(CO)_3Cl$ + $(PhCH_2)_2$	Near UV	–

1) Ref.[102]).
2) Dissapearance quantum yields for M–M species.

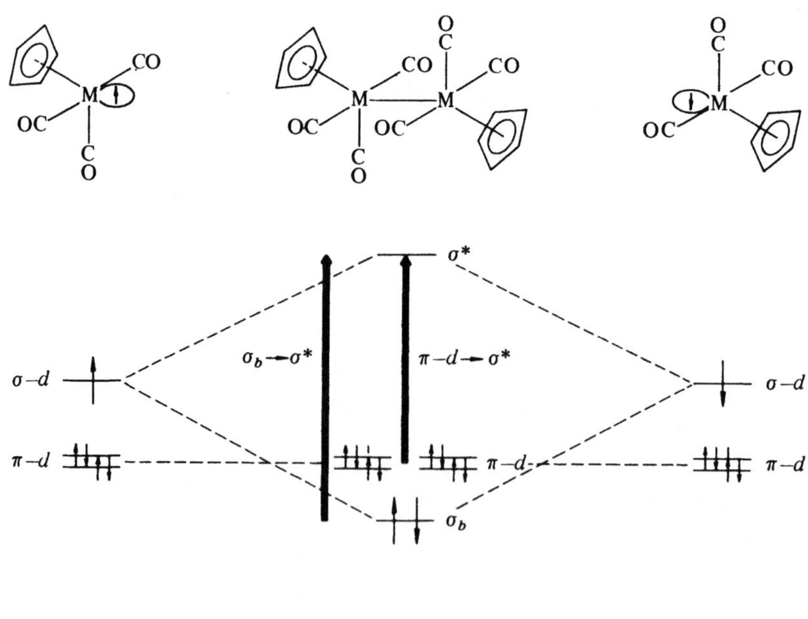

Scheme 14

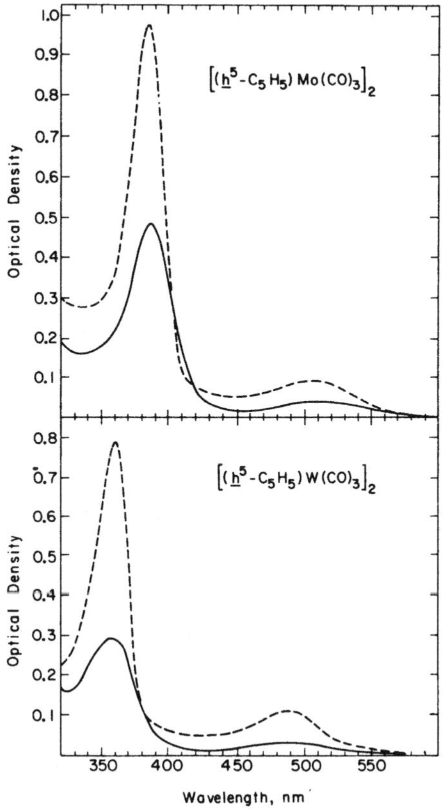

Fig. 6. Electronic absorption in EPA at 298 (——) and 77 K (---). The intense band ($\epsilon \approx 1.7 \times 10^4$) is the $\sigma_b \rightarrow \sigma^*$ (from Ref.[102])

visible absorption band near 500 nm ($\epsilon \approx 1700$ $\ell mol^{-1} cm^{-1}$) is associated with a π-$d \rightarrow \sigma^*$ transition, and the near uv band near 370 nm ($\epsilon \approx 17{,}000$ $\ell mol^{-1} cm^{-1}$) is associated with the $\sigma_b \rightarrow \sigma^*$ transition according to Scheme 14. The modest decline in quantum yields with decreasing excitation energy, Table 17, for the $[(\eta^5\text{-}C_5H_5)M(CO)_3]_2$ complexes is consistent with the electronic assignments since the π-$d \rightarrow \sigma^*$ transition should be less destructive with respect to the M—M bond than the $\sigma_b \rightarrow \sigma^*$ transition.

Finally, some data are available[103] for the d^7-d^5 heterodinuclear photoproducts in reaction (27), $(OC)_5M—M'(\eta^5\text{-}C_5H_5)(CO)_3$ (M = Mn, Re; M' = Mo, W). The electronic spectra of the four M—M' molecules are shown in Fig. 7. The spectra are seen to be related to the homodinuclear M—M and M'—M' spectra shown in Figs. 3 and 6, respectively. Despite the differences (ligands, metal, and d^n-configuration) in $M(CO)_5$ and $(\eta^5\text{-}C_5H_5)M'(CO)_3$ the electronic spectra of the M—M' complexes dictate some electronic similarity. In a simple molecular orbital scheme the similarity manifests itself by having the σ-d orbital in the mononuclear species at nearly the same energy. The low energy absorption band in the M—M' species is, thus, π-$d \rightarrow \sigma^*$ and the near uv band is again $\sigma_b \rightarrow \sigma^*$ according to Scheme 15.

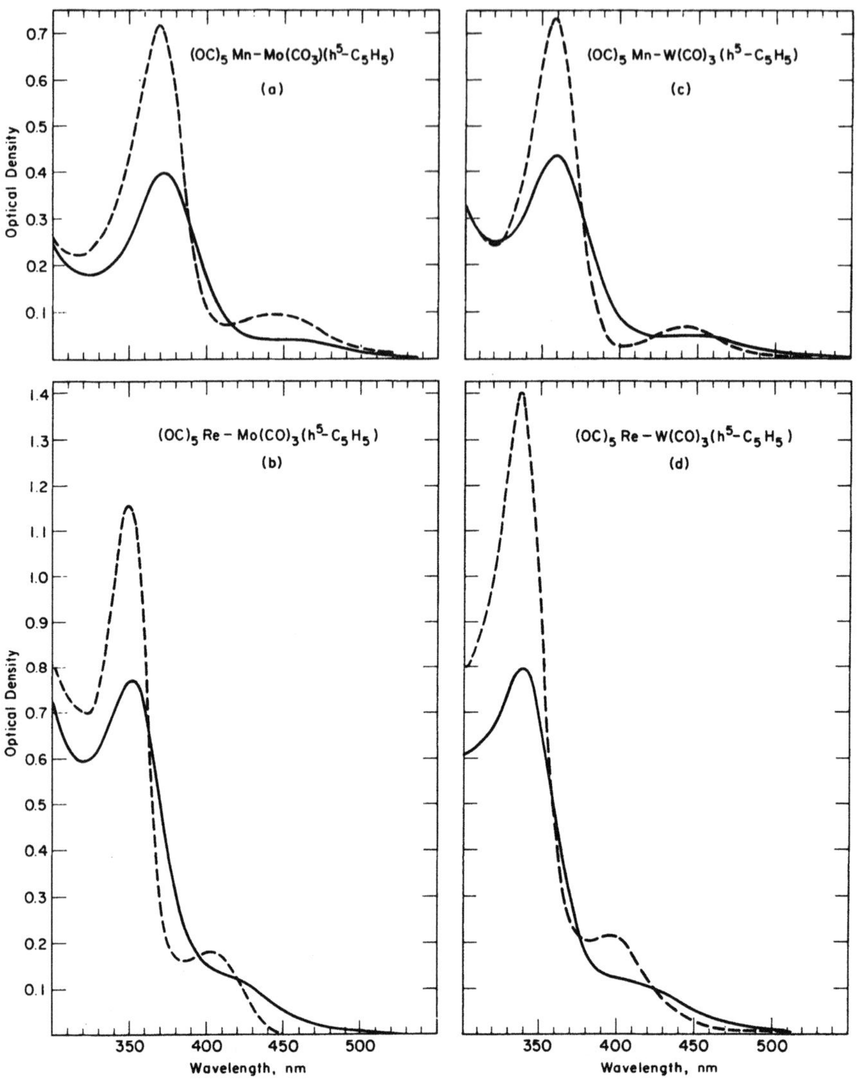

Fig. 7. Electronic absorption spectra in EPA at 298 (——) and 77 K (---). The most intense band is the $\sigma_b \rightarrow \sigma^*$ (from Ref.[103])

The photochemistry of the d^7-d^5 M–M′ molecules is also consistent with the above analogy and interpretation. First, pure samples of the M–M′ complexes undergo initial conversion to the homodinuclear M–M and M′–M′ species with high efficiency and in a ~1 : 1 ratio as expected for homolytic cleavage. Second, irradiation of the M–M′ complexes in CCl_4 yields only $M(CO)_5Cl$ and $(\eta^5\text{-}C_5H_5)M(CO)_3Cl$ with the quantum yields indicated in Table 18.[103]

The trinuclear $M_3(CO)_{12}$ (M = Fe, Ru, Os) complexes have received some attention from experimentalists, but neither the electronic spectroscopy nor the photochemistry of these complexes have been investigated in detail. The recent report[107] of the near quantitative generation of $Ru(CO)_5$ by irradiation of $Ru_3(CO)_{12}$ under CO, reaction (29), represents the type of observations that have been made, and

$$Ru_3(CO)_{12} \xrightarrow[CO]{h\nu} 3\ Ru(CO)_5 \qquad (29)$$

demonstrates the fact that declusterification is photoaccelerated. This particular example is of some interest in that it represents a novel route to the $Ru(CO)_5$. Other reactions, such as reaction (30)[106] also, show that photodeclusterification can be a route to mononuclear species. Reactions (29) and (30) likely proceed via photo-

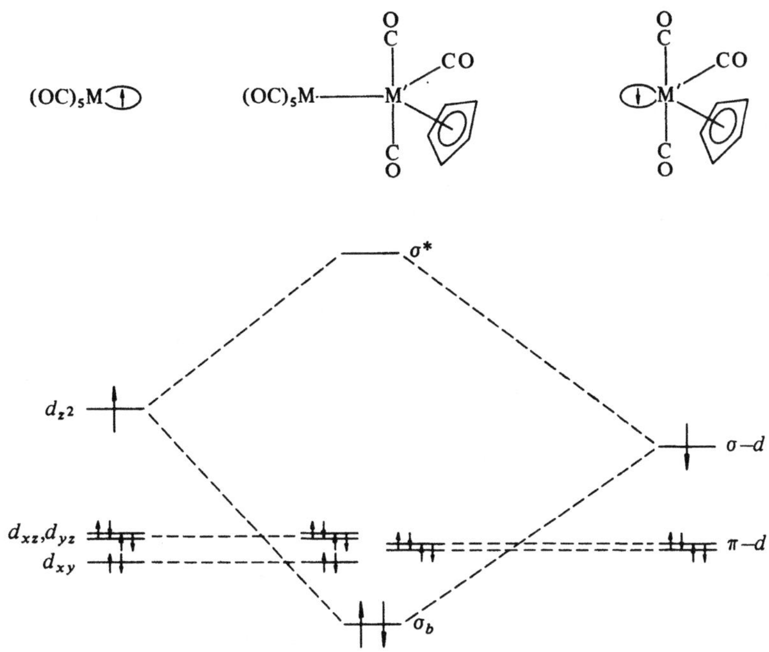

Scheme 15

Table 18. Disappearance quantum yields for $(OC)_5M{-}M'(\eta^5\text{-}C_5H_5)(CO)_3$ in CCl_4 [1])

M	M'	ϕ366 nm	ϕ436 nm
Mn	Mo	0.51	–
Re	Mo	0.56	–
Mn	W	0.35	0.05_6
Re	W	0.23	0.05_1

[1]) Ref.[103]).

$$Os_3(CO)_{12} \xrightarrow[(CH_3)_3Si\text{-}H]{h\nu} HOs(CO)_4Si(CH_3)_3 \qquad (30)$$

induced cleavage of a metal-metal bond as indicated in reaction (31). The ultimate products could arise from subsequent thermal reactions of the substrate with the

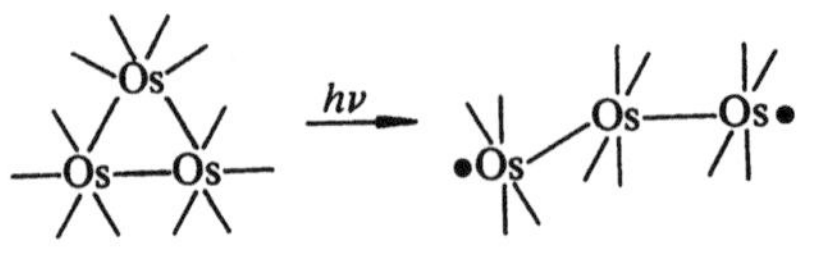

(31)

metal diradical. However, no mechanistic evidence is available to support the cleavage suggested in reaction (31), and such a postulate merits further investigation.

B. Cleavage of the Quadruple Bond in $Re_2Cl_8^{2-}$

One intriguing system which is not a metal carbonyl that has been recently investigated is $Re_2Cl_8^{2-}$.[120] Usual electronic structural formulation of the ion features a quadruple Re-Re bond,[121] and the lowest electronically excited state results from the excitation of an electron from the δ_b level to the δ^* level.[122] Higher energy electronic transitions involve electrons involved in the σ and the π Re-Re bonding.[122] Irradiation at 366 nm of $Re_2Cl_8^{2-}$ proceeds according to reaction (32).[120] The reaction does not occur upon irradiation into the low lying absorption at ~700 nm corresponding to

$$Re_2Cl_8^{2-} \xrightarrow[\substack{CH_3CN \\ \Phi \approx 10^{-2}}]{366\ nm} 2\ trans\text{-}ReCl_4(CH_3CN)_2^- \qquad (32)$$

the $\delta_b \rightarrow \delta^*$ transition. Thus, the reaction occurs from an upper electronically excited state and the fairly small quantum yields likely reflect a rapid rate of internal conversion to the lower, unreactive $\delta_b \rightarrow \delta^*$ excited state. This reaction represents the only known photoinduced cleavage of a multiple metal-metal bond. Photodissociation of multiple bonds, however, is not taboo as O_2 and CO_2 are good examples.[97] The prospects in transition metal-metal bonded systems are quite good, and investigation of polynuclear halides should prove fruitful and compliment the efforts on the organometallic cluster compounds.

III. Intraligand Photoreactions

Cases where the groups directly attached to a transition metal undergo a chemical transformation as a consequence of excited state decay of the metal complex are surprisingly difficult to find in the literature. Intraligand photoreactions, though, may well ultimately be one of the most important contributions of inorganic photochemistry. Aside from the inherently new chemistry to emerge, the hope of running energetically up-hill chemical reactions may be realized by use of intraligand photoreactions. The conversion of A to B by reactions (33)–(35) represents a metal complex photoassisted A to B conversion, and since the key step, reaction (34), is a photochemical one there is the possibility that B may be a unique chemical product

$$\text{A + Metal Complex} \longrightarrow \text{[Metal Complex-A]} \quad (33)$$

$$\text{[Metal Complex-A]} \xrightarrow{h\nu} \text{[Metal Complex-B]} \quad (34)$$

$$\text{[Metal Complex-B]} \longrightarrow \text{Metal Complex + B} \quad (35)$$

and may be thermodynamically unstable with respect to A.

Recently, a system has been reported[123] which exemplifies the essential features of the homogeneous photoassistance technique indicated in reactions (33)–(35). The *trans*→*cis*-4-styrylpyridine isomerization is up-hill energetically by ≈ 2 kcal/mole,[123] and, therefore, such an isomerization represents movement away from the thermodynamic ratio of these two isomers. Photolysis at 436 nm of $ClRe(CO)_3$(*trans*-4-styrylpyridine)$_2$ was found to ultimately yield completely (sequentially?) $ClRe(CO)_3$–(*cis*-4-styrylpyridine)$_2$. Further, the coordinated *cis*-4-styrylpyridine could be exchanged thermally with excess *trans*-4-styrylpyridine in the medium, and the sequence of events in reactions (36)–(39) was proposed[123] to account for the photoinduced iso-

$$ClRe(CO)_3(\textit{trans})_2 \xrightarrow{h\nu} ClRe(CO)_3(\textit{trans})(\textit{cis}) \quad (36)$$

$$ClRe(CO)_3(\textit{trans})(\textit{cis}) \xrightarrow{h\nu} ClRe(CO)_3(\textit{cis})_2 \quad (37)$$

$$ClRe(CO)_3(\textit{trans})(\textit{cis}) + \textit{trans} \xrightarrow{\Delta} ClRe(CO)_3(\textit{trans})_2 + \textit{cis} \quad (38)$$

$$ClRe(CO)_3(\textit{cis})_2 + \textit{trans} \xrightarrow{\Delta} ClRe(CO)_3(\textit{trans})(\textit{cis}) + \textit{cis} \quad (39)$$

merization of *trans*-4-styrylpyridine by the 436 nm irradiation of a catalytic quantity of the $ClRe(CO)_3$(*trans*-4-styrylpyridine)$_2$ complex. The *trans*-4-styrylpyridine itself has an absorption onset near 360 nm, and, therefore, does not absorb at 436 nm. The metal complex serves as a chromophore and the optical energy is dissipated into chemical paths which lead to *trans*→*cis* isomerization. For $ClRe(CO)_3$(*trans*-4-styrylpyridine)$_2$ and related complexes the lowest absorption system was found to be associated with a $\pi \to \pi^*$ intraligand (styrylpyridine) transition. The influence of the Re on the styrylpyridine electronic absorption spectrum was found to be similar to that obtained by protonating the styrylpyridine. Examples of such spectroscopic data are shown in Fig. 8. The "perturbed" intraligand excited state being lowest in energy is

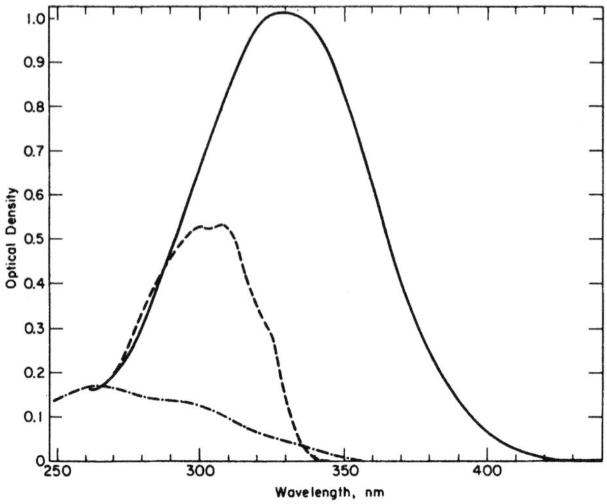

Fig. 8. Electronic absorption spectra of $ClRe(CO)_3(pyridine)_2$ (——•——•), *trans*-4-styrylpyridine (—), and $ClRe(CO)_3$(*trans*-4-styrylpyridine)$_2$ (——) all at 1.92×10^{-5} *M* in CH_2Cl_2 solution at 25 °C in 1.00 cm pathlength cells (from Ref.[123])

somewhat unique although there are some examples of 1,10-phenanthroline complexes of Rh(III) and complexes of Rh(IV) which have been shown to have lowest $\pi \rightarrow \pi^*$ states from emission experiments.[124] The red-shifted intraligand absorption of the coordinated styrylpyridine allows the photochemical isomerization to occur with lower energy light than otherwise possible.

The fact that the intraligand $\pi \rightarrow \pi^*$ transition is lowest in energy in $ClRe(CO)_3$-(*trans*-4-styrylpyridine)$_2$ also allows a direct assessment of the effect of the coordination on the decay properties of the intraligand excited state. Data for both the photoisomerization by direct irradiation, Table 19, and the triplet sensitized isomerization, Table 20, of $ClRe(CO)_3$(*trans*-4-styrylpyridine)$_2$ and *trans*-styrylpyridine

Table 19. Direct irradiation of *trans*-styrylpyridines and $XRe(CO)_3$(*trans*-styrylpyridine)$_2$ [1]

Compound	$\phi_{t \rightarrow c}$ 313 nm	$\phi_{t \rightarrow c}$ 366 nm	% *cis* at PSS ± 2 313 nm	366 nm	436 nm
trans-3-Styrylpyridine	0.48[2]	[3]	90[4]	[3]	[3]
trans-4-Styrylpyridine	0.38[5]	[3]	88[6]	[3]	[3]
$ClRe(CO)_3$(*trans*-4-styrylpyridine)$_2$	0.49	0,54	84	90	99
$BrRe(CO)_3$(*trans*-4-styrylpyridine)$_2$	0.64	0.51	99	98	>99
$ClRe(CO)_3$(*trans*-3-styrylpyridine)$_2$	0.60	0.51	93	90	99

[1] CH_2Cl_2 solutions at 25°; light intensity ~10^{-7} ein/min; ϕ's are ± 10%; Ref.[123].
[2] Literature value is 0.43, Ref. [125a], or 0.52, Ref.[125b,c].
[3] No absorption at these wavelengths.
[4] Ref.[125b,c].
[5] Literature value is 0.37, Ref.[125a], or 0.39 Ref.[125b,c].
[6] Ref.[125a].

Table 20. Triplet sensitized isomerization of *trans*-styrylpyridines and $XRe(CO)_3$ (*trans*-styrylpyridine)$_2$[1])

Compound	Sensitizer	$\phi_{t\to c}$	% *cis* at PSS
trans-4-Styrylpyridine	Benzil	0.40[2])	99
	Michler's ketone	–	67
	$Ru(bipy)_3^{2+}$	0.4 ± 0.05[3])	96.5 1.0%[3])
$ClRe(CO)_3$(*trans*-4-styrylpyridine)$_2$	Benzil	0.5	99
	Michler's ketone	–	70
	$Ru(bipy)_3^{2+}$	–	96
trans-3-Styrylpyridine	Benzil	0.44[1])	99
$ClRe(CO)_3$(*trans*-3-styrylpyridine)$_2$	Benzil	0.5	99
	$Ru(bipy)_3^{2+}$	~0.4	93

[1]) Ref.[123])
[2]) Assumed to be equal to benzophenone value, Ref.[125a]).
[3]) M. Wrighton and J. Markham, *J. Phys. Chem.*, *77*, 3042 (1973).

Table 21. Ultraviolet absorption maxima of styrylpyridine and complexes at 25 °C[1])

Compound	Solvent	Abs. max., nm (ε)
trans-3-Styrylpyridine	CH_2Cl_2	308 (19,300)
	H_2O/EtOH(9/1) pH = 7.0	304 (~19,900)
	H_2O/EtOH(9/1) pH = 2.1	296 (~19,000)
trans-4-Styrylpyridine	CH_2Cl_2	308 (27,900)
	H_2O/EtOH(9/1) pH = 7.0	306 (~28,000)
	H_2O/EtOH(9/1) pH = 2.1	338 (~28,000)
$ClRe(CO)_3$(*trans*-3-styrylpyridine)$_2$	CH_2Cl_2	297 (45,300)
$ClRe(CO)_3$(*trans*-4-styrylpyridine)$_2$	CH_2Cl_2	328 (50,300)
$BrRe(CO)_3$(*trans*-4-styrylpyridine)$_2$	CH_2Cl_2	330 (53,200)
$ClRe(CO)_3$(pyridine)$_2$	CH_2Cl_2	262 (9000)

[1]) Ref.[123]).

reveal little difference between the free and the coordinated styrylpyridine.[123]) The only definitive difference in the decay properties seems to be that there is no fluorescence from the $ClRe(CO)_3$(*trans*-styrylpyridine)$_2$ complexes while both the free *trans*-styrylpyridines and their protonated forms do fluoresce in solution at room temperature.[125]) This result may be a consequence of the fact that Re is a heavy atom and induces intersystem crossing in the coordinated ligand at a rate as to preclude radiative decay from the state reached by direct absorption.

In the $XRe(CO)_3$(*trans*-styrylpyridine)$_2$ complexes the presence of the two styrylpyridine ligands does not lead to any unusual interactions between the two organic ligands. The following facts support non-interacting styrylpyridines: (1) the molar extinction coefficients for the intraligand absorption in $XRe(CO)_3$(*trans*-styrylpyridine)$_2$ are about twice those for the free protonated ligand, Table 21, and (2) the initial isomerization quantum yields for the coordinated ligands are about the same as those for the free ligand.[123] However, in other examples of apparent examples of intraligand photochemistry involving transition metal photoassistance there seems to be some advantage in having two (or more) ligands bonded to the metal to facilitate reaction of the ligands with one another. An elegant example of the possible utility here is as in reactions (40) and (41).[126] The role of the metal may be to hold the ligand system in an orientation appropriate for cyclization. Recently,

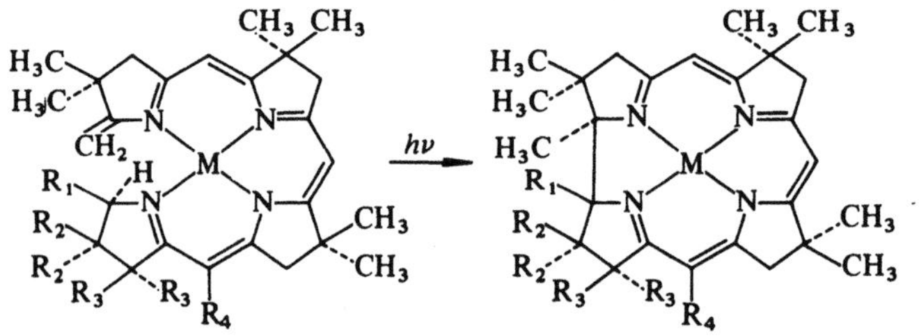

(40)

M = Zn(Cl), Mg(Cl), Cd(Cl), Pd,Pt,Li
$R_1 = R_2 = R_3 = H$
$R_4 = CN$

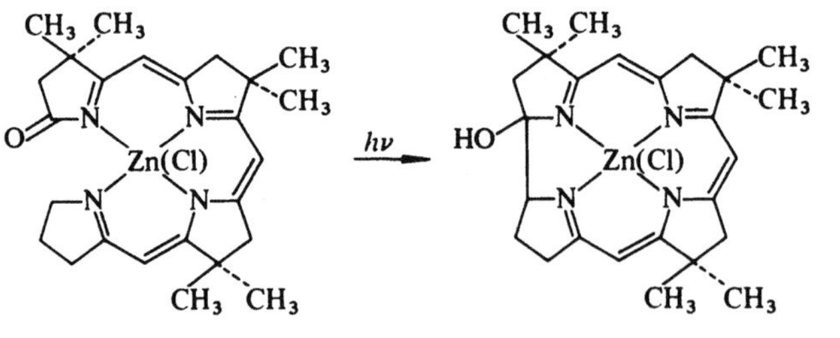

(41)

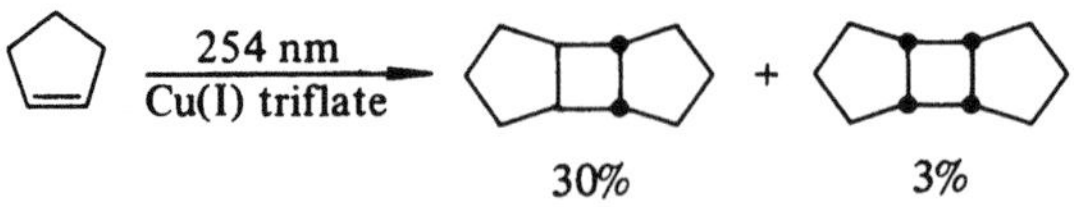

(42)

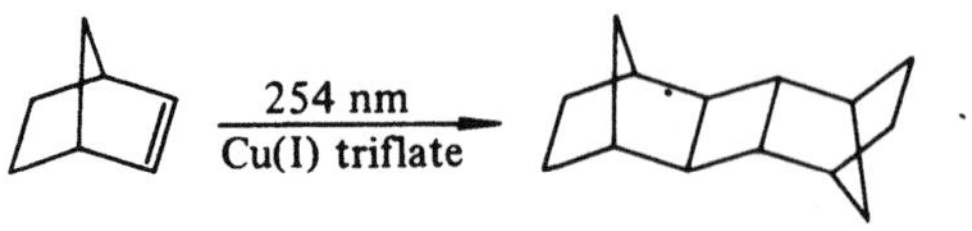

(43)

Cu(I) triflate has been used as an olefin cycloaddition photoassistance agent as in reactions (42) and (43).[127] Detailed study of reaction (43) shows that it is very likely that it is photoexcitation of a $Cu(alkene)_2^+$ complex that results in the dimerization.[127] This provides a second good example of the possible utility of transition metal photoassistance involving reactions between coordinated ligands. The nature of the excited state leading to reactions (40)–(43) is not known.

Some very interesting results have emerged from the study of several styrylpyridine complexes of Ru(II).[128] The complexes investigated include $[Ru(2,2'\text{-bipyridine})_2X_2]^{2+}$ and $[Ru(2,2'\text{-bipyridine})_2XCl]^+$ (X = *trans*-4-styrylpyridine, *cis*-4-styrylpyridine). The low energy region of the optical spectra of these complexes is dominated by Ru(II) → ligand CT absorptions, but in the region near 300 nm absorptions are observed which can be ascribed to intraligand $\pi \rightarrow \pi^*$ styrylpyridine transitions. The results of the photolysis of the four complexes are summarized in Table 22.[128] The only detectable photoreaction is styrylpyridine isomerization. The

Table 22. Styrylpyridine photoisomerization in Ru(II)-styrylpyridine complexes[1])

Complex[2])	Irrdn. λ,	$\Phi_{t \rightarrow c}$	$\Phi_{c \rightarrow t}$	% t @ PSS
$[Ru(2,2'\text{-bipy})(t\text{-}4)_2]^{2+}$	313	0.15	0.15_6	35 ± 2.3
$[Ru(2,2'\text{-bipy})(c\text{-}4)_2]^{2+}$	366	0.05	0.15	88 ± 0.3
	436	0.05	0.51	85 ± 1.5
	570	–	–	88 ± 2.3
	Michler's ketone sensitized (E_T = 61 kcal/mole)	0.03	0.04	65 ± 2.1
	Zn etioporphyrin I (E_T = 41 kcal/mole	0.001	0.02	96 ± 2.9
	Etioporphyrin (E_T = 40 kcal/mole)	–	–	93 ± 1.0
$[Ru(2,2'\text{-bipy})(t\text{-}4)Cl]^+$	313	0.09_2	0.12_3	23.6 ± 1.1
or				
$[Ru(2,2'\text{-bipy})(c\text{-}4)Cl]^+$	436	0.03_5	0.665	98.6 ± 0.9

1) Ref.[128].
2) All results at 25° in butyronitrile.

data show that there is a strong wavelength dependence on the photoisomerization activity which is interpreted as intraligand excited state activity at 313 nm excitation and mainly Ru(II)→ligand CT excited state reactivity at the lower excitation energies. The fact that the intraligand excited state shows different reactivity patterns than the CT states is consistent with the nature of the one electron transitions in each case. In the intraligand $\pi \rightarrow \pi^*$ transition respectable quantum yields for both *trans→cis* and *cis→trans* isomerization obtain as found in the free ligand. Upon irradiation into the Ru(II)→styrylpyridine CT transition the resulting excited state could be viewed as a radical anion of styrylpyridine coordinated to Ru(III). The coordinated radical anion of styrylpyridine apparently results in very effective *cis→trans* isomerisation, but little *trans*→*cis* isomerization as expected from arguments concerning the thermodynamic properties of *cis*- and *trans*-styrylpyridine. The studies of the Ru(II) complexes show that the CT excited states do have particular reactivity properties, but, also, the results show that intraligand excited state reactions can compete with internal conversion to lower excited states. Chemists may take advantage of both of these conclusions on the one hand to discover new chemistry associated with the CT transitions and on the other hand to tap intraligand photochemistry modified by coordination to a metal as in reactions (40) and (41).

Styrylpyridine photochemistry has been important in one other system. The photoprocesses in the complexes $W(CO)_5X$ (X = pyridine, 2-styrylpyridine, and 4-styrylpyridine) have been investigated.[129] Unlike the $XRe(CO)_3$(*trans*-styryl-pyridine)$_2$ and Ru(II)-styrylpyridine complexes having lowest IL and CT states, respectively, the $W(CO)_5X$ complexes have lowest LF excited states with only a small contribution from W → pyridyl CT. The one-electron diagram for low-spin d^6, C_{4v} complexes shown in Scheme 7 is appropriate here. Both photosubstitution and photo-isomerization reactions are found for the $W(CO)_5X$ complexes and some quantum yield data are found in Table 23.[129] The data show that substitution efficiencies for

Table 23. Quantum efficiencies for photoreactions of $W(CO)_5(X)$ complexes[1)]

Reactions

A. $W(CO)_5$(pyridine) $\xrightarrow{h\nu}$ $W(CO)_5$(l-pentene)

B. $W(CO)_5$(pyridine) $\xrightarrow{h\nu}$ *cis*-$W(CO)_4$(pyridine)$_2$

C. $W(CO)_5$(*t*-4-styrylpyridine) $\xrightarrow{h\nu}$ $W(CO)_5$(l-pentene)

D. $W(CO)_5$(*t*-4-styrylpyridine) $\xrightarrow{h\nu}$ $W(CO)_5$(*c*-4-styrylpyridine)

E. $W(CO)_5$(*t*-2-styrylpyridine) $\xrightarrow{h\nu}$ $W(CO)_5$(*c*-2-styrylpyridine)

F. $W(CO)_5$(*t*-2-styrylpyridine) $\xrightarrow{h\nu}$ $W(CO)_5$(1-pentene)

Reaction	$\phi_{436\ nm}$	$\phi_{366\ nm}$	$\phi_{313\ nm}$	$\phi_{254\ nm}$
A[2)]	0.63	0.50	0.38	0.3$_4$
B[3)]	0.00$_2$	0.01$_3$	0.03$_9$	0.04
C[2)]	0.16	0.08	0.05	–
D	0.49	0.34	0.26	0.21
E	0.08 ($t \rightarrow c$) 0.31 ($c \rightarrow t$)	–	–	–
F[2)]	0.16	0.07	0.13	–

1) Ref.[129].
2) ϕ measured at room temperature in the presence of 3.66 *M* 1-pentene, isooctane solvent.
3) ϕ for formation of *cis*-$W(CO)_4(py)_2$ at room temperature in the presence of 0.25 *M* pyridine in isooctane.

the styrylpyridine complexes are smaller than for the pyridine complex, but both the quantum yield for isomerization and the substitution processes in the styrylpyridine complexes have the same trends as a function of excitation wavelength as the pyridine photosubstitution quantum yields. These data suggest than the LF states can decay to a state responsible for the *cis-trans* isomerization phenomena. The notion that the LF excitation energy can be dissipated into intraligand reaction pathways merits further investigation. Obviously, *trans*→*cis*-styrylpyridine isomerization does take place upon excitation into the lowest absorption band ($\lambda_{max} \approx 410$ nm) in $W(CO)_5$ (*trans*-styrylpyridine) in contrast to the situation with the Ru(II)-*trans*-styrylpyridine complexes. By way of contrast to $W(CO)_5$(*trans*-4-styrylpyridine), initial data for *cis*-$W(CO)_4$-(*trans*-4-styrylpyridine)$_2$ show that the lowest excited state has substantial contribution from W→*trans*-styrylpyridine CT and shows extremely inefficient *trans*→*cis*-4-styrylpyridine photoisomerization upon excitation into the lowest absorption system.[130)]

Despite their relative insensitivity to light d^6 metallocenes have received considerable attention with respect to intraligand photoreactions. Most studies seem to have begun with the notion that the metallocenyl unit is analogous to an organic aromatic unit but with low energy LF and CT absorptions. Ferrocene and its derivatives have received the most study. In the absence of any good electron acceptors ferrocene itself is essentially inert to irradiation in the optical region. The photochemical Fries, reaction (44), seems to be one of the first reactions of this type.[131)]

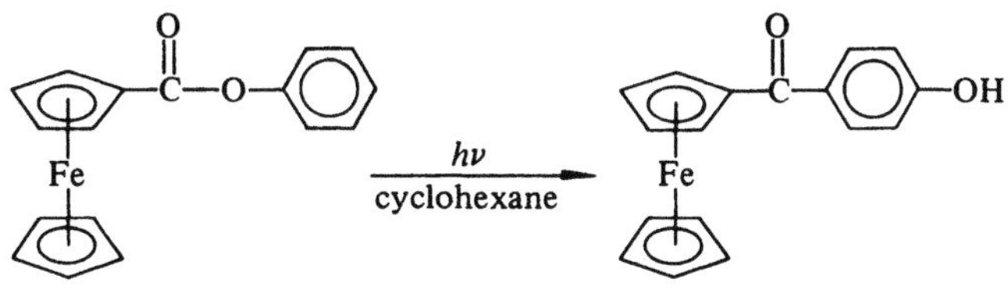

(44)

Reaction (45) is a more recent example of a photoinduced ligand reaction of

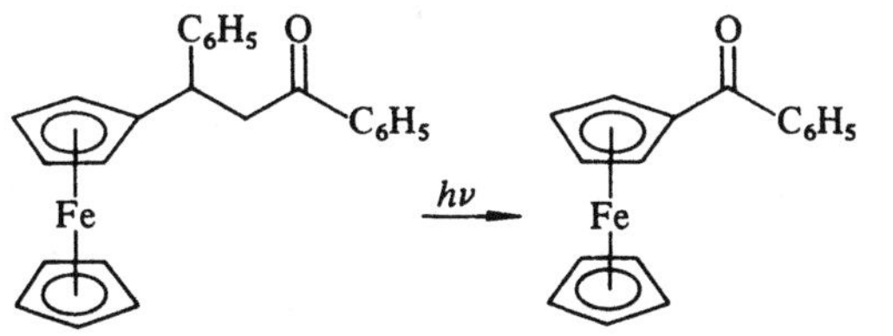

(45)

a ferrocene.[132)] Both reaction (44) and (45) likely proceed via upper excited states involving principally IL transitions. Attempted photoreduction of benzoylferrocene in isopropanol by analogy to benzophenone has failed.[133)] In this case the formation of an IL $n \rightarrow \pi^*$ state either does not obtain or the lifetime is too short to allow intermolecular reactions with the solvent. Continued study of these and related systems should focus on an attempt to understand the internal conversion processes in these molecules to determine what factors control whether an intraligand excited state can give efficient *chemical* decay.

The study of the direct irradiation and triplet sensitized isomerization of styrylferrocene, reaction (46), is noteworthy.[134)] For either direct or benzophenone sensitized photoisomerization the photostationary state is exclusively the *trans* isomer.

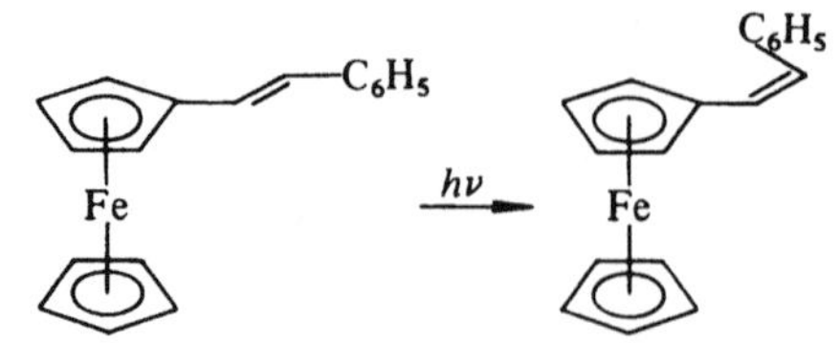

(46)

However, while 313 and 366 nm direct irradiation yields a *cis→trans* quantum yield of ~0.005, the benzophenone triplet sensitized and the low energy direct irradiation yield is only ~0.001. These results support the notion that two reactive excited states are involved: one associated with the styrylferrocene which leads to the more efficient isomerization and a ferrocene-type excited state at lower energy which has a smaller tendency to undergo *cis→trans* isomerization. It is interesting to note that the 100% *trans* photostationary state is likely very close to the amount expected at thermal equilibrium. Such a result would be consistent with an intramolecular CT transition to yield a radical-anion ferricinium complex as in reaction (47). The radical anion could yield *cis→trans* isomerization just as in the Ru(II)-styrylpyridine complexes. The wavelength dependence of the isomerization of the styrylferrocene is consistent with (47) in that upper excited states are required for the photooxidation of ferrocene to ferricinium in CCl_4.[135)]

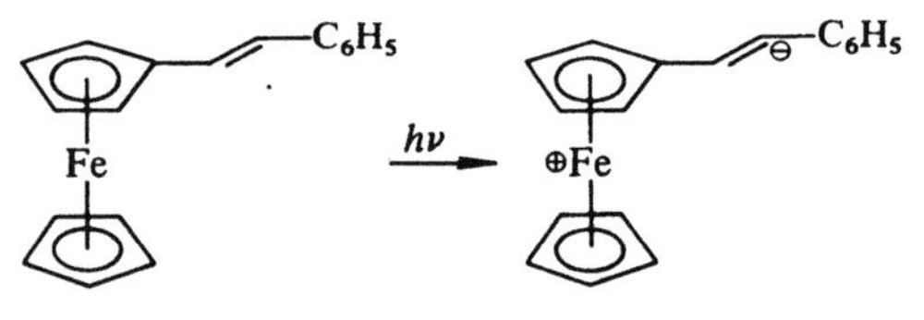

(47)

Aside from the Cu(I) photoassisted bimolecular alkene cycloaddition reactions such as (42) and (43)[127)] there are a number of examples of apparent transition metal photoassisted reactions of olefins[2c)] where the olefin is bonded to the metal through the olefin π-system. Olefin reactions such as *cis-trans* isomerization,[2c, 136)] 1,3-hydrogen shifts,[2c, 137)] dimerization, hydrosilation,[139)] and hydrogenation[140)] have been studied to some extent. In the systems investigated thus far, the role of the light has been either to lead to olefin reaction promptly or to generate a very reactive intermediate leading to olefin reaction by a series of secondary thermal reactions. Detailed mechanistic information is generally unavailable as in many cases the thermal lability of the olefin complexes precludes detailed knowledge concerning the stoichiometry of the species actually undergoing photoexcitation.

One of the cases where the role of the light is to produce reaction directly is reaction (48).[141)] This reaction does lend itself to further detailed study to attempt

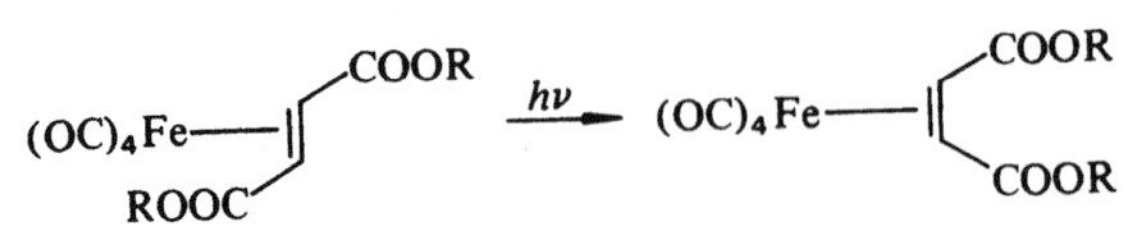

(48)

to establish the nature of the intermediate responsible for the isomerization of the ligand. One proposal is that electronic excitation produces a CT state such that a coordinated olefin radical anion is generated which can decay to either the *cis* or *trans* olefin.[2c)] The production of a free olefin in an electronically excited state which can decay to either the *cis* or *trans* isomer is seemingly improbable but easily tested.

The conversion of *trans*- to *cis*-stilbene by irradiation of either $W(CO)_6$ or $Mo(CO)_6$ is consistent with production of an excited metal carbonyl-stilbene complex which undergoes isomerization promptly.[136, 137)] The generation of the *cis* isomer represents movement away from the thermodynamic ratio of *cis*- and *trans*-stilbene which is ~100% *trans*. Thus, an electronically excited state must be important in the reaction sequence. Two proposals have been made:[137)] (1) excitation of an $M(CO)_n$ $(stilbene)_{6-n}$ complex leads to a sigma bonded diradical intermediate as shown in reaction (49) or (2) electronic excitation may yield a perturbed IL excited state as for the styrylpyridine complexes discussed above. The sigma bonded olefin diradical

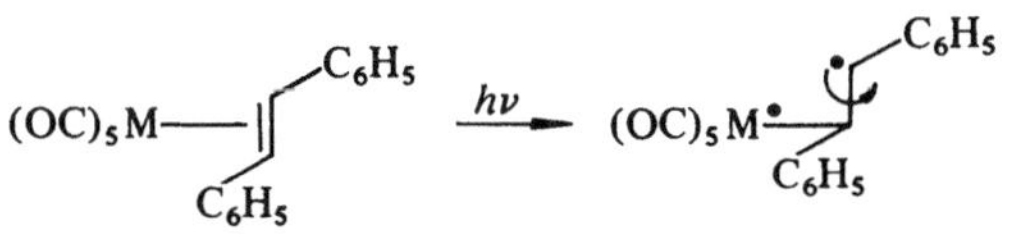

(49)

intermediate has some appeal as it might be viewed as arising from the interaction of the olefin with electronically excited $W(CO)_5$ species having an electron in a strongly σ^* orbital, say d_{z^2}. The $Co(CN)_5^{3-}$ species in its ground electronic state has an electron in the d_{z^2} orbital,[142)] and the $Co(CN)_5^{3-}$ species reacts with acetylene, reaction (50), and other olefins.[143)] This seems to be a reasonable analogy to the reaction of

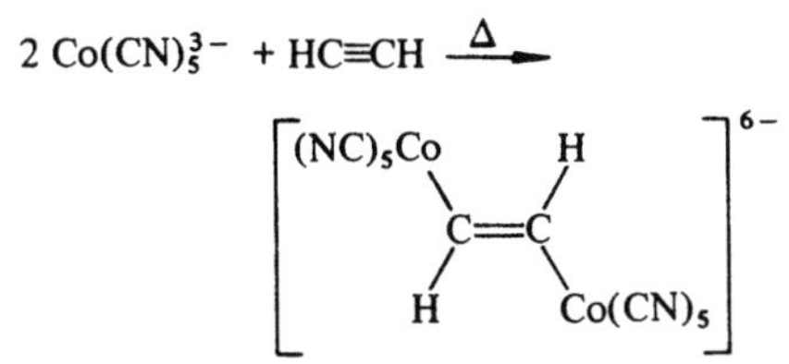

(50)

electronically excited $W(CO)_5$ and an olefin. The proposal indicated in reaction (49) gains some credibility from the observation that 1,2-dideuterioethylene undergoes *trans*→*cis* isomerization upon irradiation of $W(CO)_6$ in solutions of 1,2-dideuterioethylene.[137)] The ethylene intraligand excited state is seemingly too high energetically to be of importance in this system. It should be pointed out that an intermediate similar to that in reaction (49) could also be invoked to explain the isomerization indicated in reaction (48).

The $M(CO)_6$ (M = Mo, W) photoassisted interconversion of the linear pentenes, reaction (51), is an example of a situation where the role of the light, at least in part, is to generate a reactive intermediate which is responsible for the isomerization reactions.[137)] The key photoreaction is dissociative loss of CO from $W(CO)_5$(alkene), reaction (52), to yield a coordinatively unsaturated intermediate which can lead to

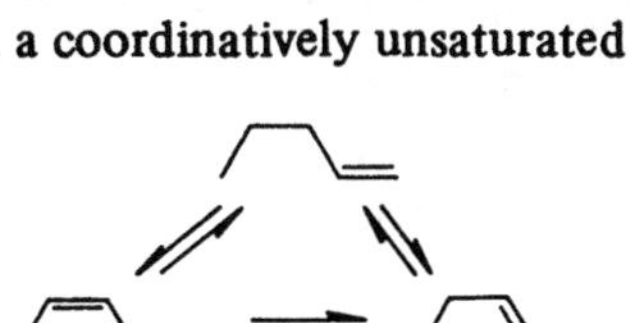

(51)

$$W(CO)_5(\text{alkene}) \xrightarrow{h\nu} W(CO)_4(\text{alkene}) + CO \quad (52)$$

sustained alkene isomerization *without* irradiation by the sequence of reactions (53) through (57). The sustained pentene isomerization without light takes place

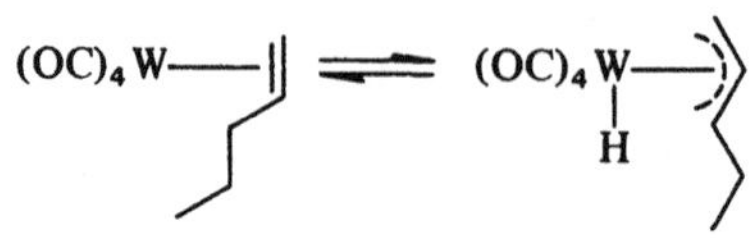

(53)

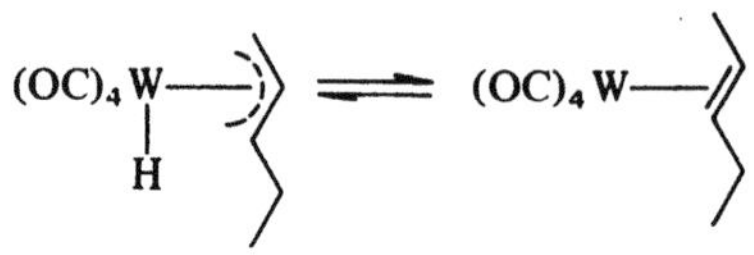

(54)

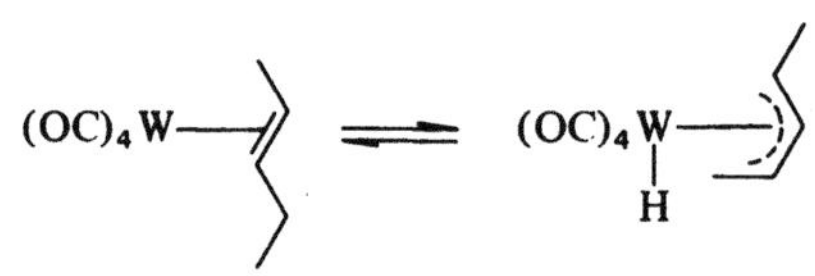

(55)

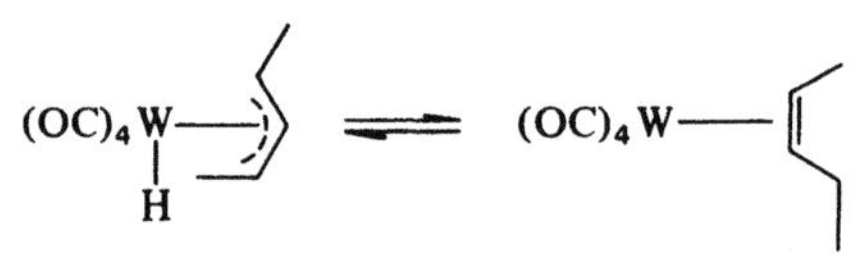

(56)

$$(OC)_4W\,(\text{pentene}) \underset{-\text{pentene}}{\overset{+\text{pentene}}{\rightleftharpoons}} (OC)_4W\,(\text{pentene})_2 \quad (57)$$

due to the thermal lability of the $W(CO)_4(\text{pentene})_2$. This situation is one appropriately termed *photocatalysis* since a catalyst has been photogenerated which can operate thermally without additional photolysis: the photons themselves can be used catalytically simply to initiate the catalysis. For the case at hand there is likely a component of the isomerization proceeding by the intermediate indicated in reaction (49), and this results in higher sustained rates of isomerization under constant irradiation beyond the thermal rate in the dark. Also, it is to be noted that, in general, photoexcitation can accelerate thermal processes such as ligand exchange, and consequently, continuous irradiation can be used to accelerate thermal catalysis.

Olefin isomerization has also been mediated by the photolysis of $Fe(CO)_5$.[144)] Recently, a detailed study of alkene isomerization by photolysis of $Fe(CO)_5$ has shown that the reaction is truly photocatalytic.[145)] The very high quantum yields (>> 1.0), Table 24, and the fact that the pentenes are ultimately equilibrated to the thermodynamic ratio support the notion that the role of the light is to generate a thermally active catalyst. A mechanism similar to that in reactions (53)–(57) involving $Fe(CO)_3$ as the repeating unit can be used to account for the results.

The $M(CO)_6$ (M = Cr, Mo, W) photocatalyzed hydrogenation[140)] and hydrosilation[139)] of 1,3-dienes is another example where the role of the light is to generate coordinative unsaturation. The reactions (58) and (59) can be carried out on a synthetic scale and represent very selective transformations. The facts that

Table 24. $Fe(CO)_5$ photocatalyzed isomerization of l-pentene to 2-pentene[1])

l-Pentene, *M*	ϕ (Isomerization)
0.04	1.1
0.10	1.5
0.5	14.5
2.0	68
9.14	429

[1]) Ref.[145]).

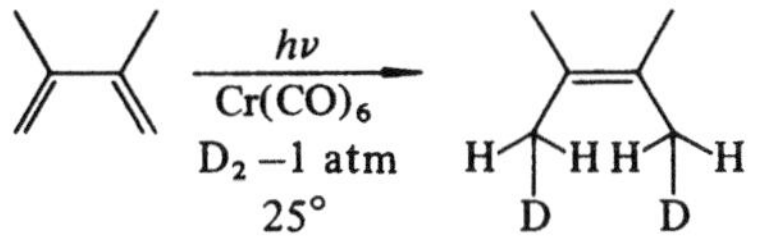

(58)

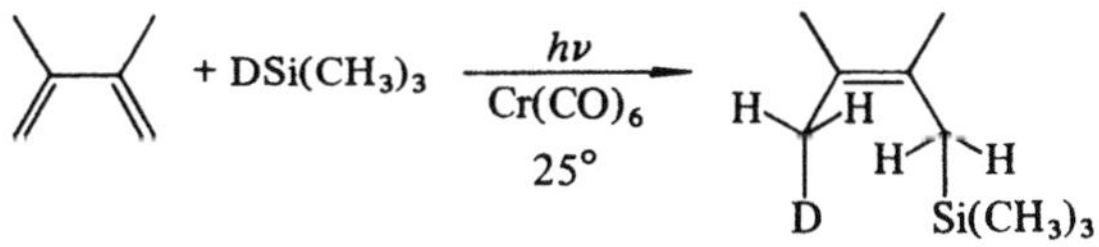

(59)

(1) (arene)$Cr(CO)_3$ catalyzes both of these reactions in the dark at high temperatures[146]) and (2) that arene group exchange is a known[147]) thermal process in (arene)$Cr(CO)_3$ suggest that $Cr(CO)_3$ is the repeating unit in reactions (58) and (59). The substitutionally labile (with respect to CH_3CN substitution) $(CH_3CN)_3Cr(CO)_3$ complex catalyzes reaction (58) at 40 °C, providing further support for the notion that $Cr(CO)_3$ is the repeating unit in the photoreactions.[148]) Illumination of (benzene)$Cr(CO)_3$ does not yield the photocatalyzed hydrogenation reaction[149]) and this is likely due to the fact that photolysis of (benzene)$Cr(CO)_3$ yields photoinduced loss of CO, reaction (60), and not loss of the arene group.[150])

$$\text{(benzene)}Cr(CO)_3 \xrightarrow{h\nu} \text{(benzene)}Cr(CO)_2 + CO \qquad (60)$$

Several interesting examples of intraligand chemistry involving oxidation or reduction of "inorganic ligands" will be discussed in the final section of this paper, but a discussion of the photochemistry of azide complexes involving the photogeneration of coordinated nitrenes is appropriate here. The irradiation of $Ir(NH_3)_5N_3^{2+}$ proceeds according to reaction (61) in the presence of 0.1 *M* HCl.[151]) The quantum efficiency of reaction (61) for irradiation in the range 250–400 nm

$$Ir(NH_3)_5N_3^{2+} + 2\,H^+ + Cl^- \xrightarrow{h\nu} Ir(NH_3)_5NH_2Cl^{3+} + N_2 \qquad (61)$$

is essentially invariant at nearly 0.6. The products are certainly consistent with the primary photogeneration of $Ir(NH_3)_5N^{2+}$, and importantly, though LF excited states are in the energetic range used for excitation, no simple ligand substitution chemistry was observed. The photogeneration of the nitrene intermediate, then, may occur from a low lying azide localized excited state which is below the LF excited states, but this reactive azide excited state does not give rise to an observable absorption band.

The photochemistry[152)] of $Rh(NH_3)_5N_3^{2+}$ also seems to have a component of a nitrene generation pathway as in reaction (61), but at higher excitation energies CT states are populated which decay to give azido radicals and presumably $Rh(NH_3)_5^{2+}$ species. A comparison of quantum yields for the $M(NH_3)_5N_3^{2+}$ (M = Co, Rh, Ir) complexes are given in Table 25.[151–153)] The $Co(NH_3)_5N_3^{2+}$ gives no

Table 25. Efficiency of photoreactions of $M(NH_3)_5N_3^{2+}$

Complex	Irrdn. λ, nm	ϕ_{decomp}	$\phi_{nitrene\ formn.}$	Ref.
$Rh(NH_3)_5N_3^{2+}$	250	0.31 ± 0.01	0.18 ± 0.01	152)
	275	0.178 ± 0.006	0.096 ± 0.005	
	300	0.104 ± 0.004	0.053 ± 0.002	
	320	0.054 ± 0.003	0.043 ± 0.002	
	350	0.035 ± 0.002	0.034 ± 0.002	
$Ir(NH_3)_5N_3^{2+}$	250	0.63	0.63	151)
	300	0.61	0.61	
	350	0.83	0.83	
	400	0.65	0.65	
$Co(NH_3)_5N_3^{2+}$	254	~0.5	None	153)
	550	0.011	None	

nitrene formation[153c)] and apparently represents the extreme in CT photochemistry whereas $Ir(NH_3)_5N_3^{2+}$ having the highest energy CT ($N_3^- \rightarrow M$) absorption gives exceedingly high yields for nitrene formation via an azide localized excited state. The Rh complex is clearly intermediate in behavior. Recently, Zink[154)] has reported molecular orbital calculations that are consistent with the IL excited state reactivity proposal based on the experimental work.[151, 152)] Zink makes the additional interesting suggestion that photolysis of $Rh(NH_3)_5N_3^{2+}$ could lead to $Rh(NH_3)_5N_2^{3+}$ followed by rapid thermal aquation to yield $Rh(NH_3)_5OH_2^{3+}$. The formation of $Rh(NH_3)_5OH_2^{3+}$ is observed and has been ascribed[153c)] to formation of $Rh(NH_3)_5^{2+}$ and N_3 followed by an oxidation of the Rh^{2+} species and ultimate formation of the $Rh(NH_3)_5OH_2^{3+}$. Other examples where coordinated nitrenes seem to result from photolysis are shown in reactions (62) and (63).[155)]

$$(\eta^5\text{-}C_5H_5)Mo(CO)(PPh_3)_2(N_3) \xrightarrow{h\nu}$$

$$(\eta^5\text{-}C_5H_5)Mo(CO)_2(PPh_3)(NCO) + (\eta^5\text{-}C_5H_5)Mo(CO)(PPh_3)_2(NCO) \quad (62)$$

$$(\eta^5\text{-}C_5H_5)Mo(CO)(NO)(PPh_3) \xrightarrow{h\nu}$$

$$(\eta^5\text{-}C_5H_5)Mo(CO)_2(NCO)(PPh_3) + (\eta^5\text{-}C_5H_5)Mo(CO)(NCO)(PPh_3)_2 \quad (63)$$

IV. Redox Reactions and Charge-Transfer State Chemistry

Primary photoreactions leading to net oxidation or reduction reactions of coordination compounds are well known and are often the result of decay paths accessible only from CT states. A number of coordination compounds yield photoelectron production in solution, the Ru(2,2′-bipyridine)$_3^{2+}$ ion has been shown to be an electron donor from its electronically excited state, and photoreduction of several metal complexes has been studied in detail. Discussion of these three areas should reveal most of the important principles associated with photoredox and CT state chemistry.

A. Ru(2,2′-bipyridine)$_3^{2+}$

Interest in Ru(2,2′-bipyridine)$_3^{2+}$ has stemmed from the early observations[156] that the ion luminesces in fluid solution at room temperature subsequent to electronic excitation. The number of coordination complexes that luminesce under conditions where photochemistry occurs is very small, but includes a few Cr(III) complexes which exhibit LF emission,[157] XRe(CO)$_3$(1,10-phenanthroline) (X = Cl, Br, I) and related complexes which emit from a state which is Re → 1,10-phenanthroline in absorption,[158] and Ir(2,2′-bipyridine)$_3^{3+}$ and Ir(1,10-phenanthroline)$_3^{3+}$ from which emission is observed.[159] Study of Ru(2,2′-bipyridine)$_3^{2+}$ has been particularly intense. The emission is associated with an excited state which is Ru(II)→2,2′-bipyridine CT in absorption. The spin multiplicity of the emitting state is logically triplet, but spin-orbital coupling in Ru(II) is great enough as to preclude assignment of discrete spin-states.[160] The temperature dependence of the emission lifetime of Ru(2,2-bipyridine)$_3^{2+}$ has been studied in detail and interpreted in terms of a thermally equilibrated manifold of spin-orbit states which each can decay to the ground state with a different set of radiative and nonradiative decay constants.[161] The energetic splitting of this manifold of states arises from the spin-orbital coupling perturbation. The energy level diagram for Ru(2,2′-bipyridine)$_3^{2+}$ is shown in Scheme 16. Other Ru(II) complexes including Ru(4,4′-diphenyl-2,2′-bipyridine)$_3^{2+}$ and *cis*-Ru(CN)$_2$(2,2′-bipyridine)$_2$ have been studied in the

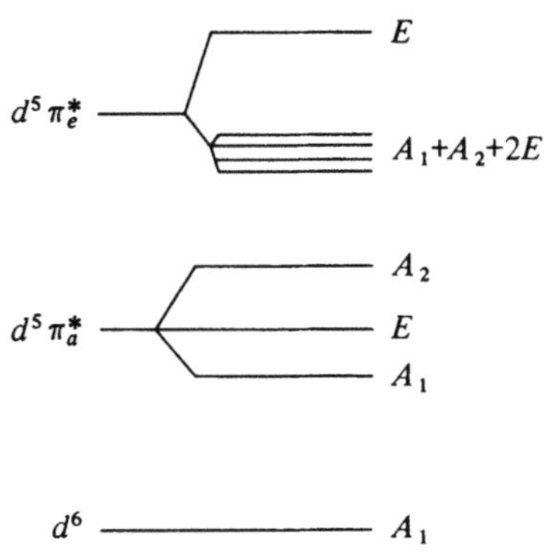

Ru(2,2′–bipyridine)$_3^{2+}$ Energy Levels[161]

Scheme 16

same way and fitted to the spin-orbital coupling model.[161] Since the energetic spacings of the spin-orbit states are fairly small, it is not clear what importance the data will have for an understanding of the chemistry to be found. But, the complete understanding of these systems should contribute significantly to the molecular basis for decay processes.

The fairly long solution lifetime of the charge-transfer excited state of Ru(2,2'-bipyridine)$_3^{2+}$ has been exploited by using it as a "triplet" sensitizer for both organic[162] and inorganic reactions.[163] Interest in the photochemistry associated with the charge-transfer state was stimulated by the data shown in Table 26[164] which led Gafney and Adamson to propose that Ru(2,2'-bipyridine)$_3^{2+}$ could be used as an *excited state reducing agent.*[164] The fact that the excited Ru(II) complex can serve as a reducing agent is now well established and several reactions have been studied. Substrates that have been reduced by the excited Ru(II) complexes are given in Table 27.[164–168] It is interesting to note here that under certain circumstances reduction of Ru(2,2'-bipyridine)$_3^{3+}$ yields electronically excited Ru(2,2'-bipyridine)$_3^{2+}$ which undergoes its characteristic decay paths including luminescence.[169]

Table 26. Ru(bipy)$_3^{2+}$ excited state reduction of $Co(NH_3)_5X^{2+}$ [1]

Complex	$\phi_{Co^{2+}}$ formn.[2]	K_{sv},[3] M^{-1}
$Co(NH_3)_6^{3+}$	Small	–
$Co(NH_3)_5F^{2+}$	0.003	0
$Co(NH_3)_5Cl^{2+}$	0.063	100
$Co(NH_3)_5Br^{2+}$	0.104	225

[1] Ref.[164].
[2] Determined for 1 x 10^{-3} *M* complex.
[3] Stern-Volmer quenching constant.

Table 27. Substrates reduced by electronically excited Ru(2,2'-bipyridine)$_3^{2+}$

Substrate	Ref.
$Fe(OH_2)_6^{3+}$	165)
Co(III)-ammines	166, 164)
Tl^{3+}	167)
[1,1'-Dimethyl-4,4'-bipyridine]$^{2+}$	165)
[*trans*-1,2-Bis(N-methyl-4-pyridyl)ethylene]$^{2+}$	165)
$Ru(NH_3)_6^{3+}$	165, 166)
$Ru(NH_3)_5Cl^{2+}$	166)
$Co(C_2O_4)_3^{3-}$	168)

The principle established with the Ru(2,2'-bipyridine)$_3^{2+}$ complex is that quenching of the excited state can occur by electron transfer when such a redox process is kinetically and energetically favorable. Other examples of this phenomenon are likely to follow.

B. Photoelectron Production from Transition Metal Complexes

Photoelectron production in solution has been verified or suspected from a number of metal complexes. A collection of metal ions that form complexes which likely yield photoelectrons upon irradiation includes Cr(II), Mo(IV), W(IV), Fe(II), and Ru(II).[2b)] It is readily appreciated that a photoelectron producing species must have available an oxidation state one level above the starting level and one which is energetically accessible with light of optical energies.

Among the several examples of photoelectron producing complexes, Fe(CN)$_6^{4-}$ has been studied in greatest detail.[170–177)] The quantum yield for photoelectron production is strongly wavelength dependent, Table 28,[177)] and shows definitively

Table 28. Wavelength dependence for photoelectron production from Fe(CN)$_6^{4-}$ [1)]

Irrdn. λ, nm	ϕ [2)]
214	0.88
228	0.89
253.7	0.65
265	0.52
289	0.18
303	0.14
313	0.104
365	<0.02

[1)] Ref.[177)].
[2)] Actually this is the quantum yield for N_2 formation from N_2O scavenging of the photoelectrons that escape primary geminate recombination.

that the lowest excited states do not lead to substantial yields of the photoelectrons. The lowest states are the LF $^{1,3}T_{1g}$ and $^3T_{2g}$ (*cf.* Table 5) and these apparently lead to very efficient substitution.[64, 178)] Redox reaction at high energy and substitution at low energy is not unusual, but for both of these to be efficient is rather unique.

Full details subsequent to high energy electronic excitation of Fe(CN)$_6^{4-}$ have not been elucidated, but some data do support the notion that a Fe(CN)$_6^{4-}$ → solvent CT (CTTS) excited state is responsible for the ejection of the electron. Shirom and Stein[177)] have postulated the reasonable mechanism indicated in Scheme 17. Despite the fact that there appears to be an absorption associated with the CTTS transition there is no necessity to invoke direct population of the CTTS excited species as it could be populated by relaxation of the $^1T_{1u}$ excited Fe(II)→CN$^-$ CT state.

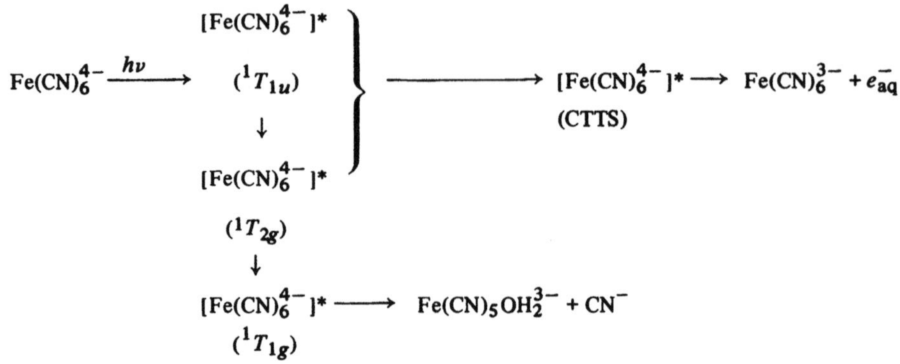

Scheme 17

C. Primary Photoredox Reactions of Inorganic Systems

Aside from the chemically clean redox reactions described above involving the electronically excited Ru(2,2′-bipyridine)$_3^{2+}$ and the dissociation of electrons from excited species there are a large number of photoinduced redox reactions of metal complexes which involve more extensive reorganization of the coordination sphere. There are, in fact, only a small number of photochemically induced redox reactions which yield products differing only in oxidation state. Production of photoelectrons provides a good pathway for photooxidation, but few, if any, examples exist for photoinduced reduction *not* involving extensive changes in coordination sphere. Irradiation of $M(CN)_8^{3-}$ (M = Mo, W) is observed[179–180] to yield $M(CN)_8^{4-}$ which is one apparent photoreduction not involving anything other than reduction in the photoreceptor. Obviously, the reduction must be accompanied by an oxidation as indicated in reaction (64) where the solvent H_2O is oxidized. Since $M(CN)_8^{4-}$ itself yields photoelectrons[2b] upon high energy excitation one could conceivably reduce and oxidize water by irradiation of a mixture of $M(CN)_8^{3-}$ and $M(CN)_8^{4-}$ as has been proposed[181] for certain transition metal aquo ions.

$$4\,M(CN)_8^{3-} + 2\,H_2O \xrightarrow{h\nu} 4\,M(CN)_8^{4-} + 4\,H^+ + O_2 \qquad (64)$$

A well studied example of a photoreduction which is accompanied by decomposition is the CT state photochemistry of Co(III)-ammine complexes.[2b] In recent years substantial effort[182, 183, 2b] has been directed towards understanding the photoreductions like reaction (65). Reactions of this type almost certainly origi-

$$Co(NH_3)_5Br^{2+} \xrightarrow{h\nu} Co(NH_3)_5^{2+} + \cdot Br \qquad (65)$$

nates from CT states of the $X^- \rightarrow$ Co(III) CT type. The resulting excited state can be viewed in simplest terms as a Co(II) complex which is quite substitutionally labile. Actually, it is possible to use simple one-electron level diagrams like those used in discussing the M–M bonded complexes. Say, for $Co(NH_3)_5Br^{2+}$ the appropriate diagram is shown in Scheme 18. Now, the $\sigma_b(p_z) \rightarrow \sigma^*\,(d_{z^2})$ one-electron transition is

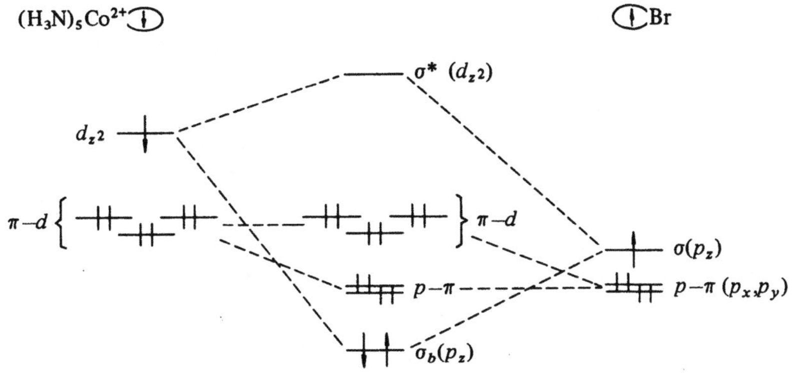

Scheme 18

the σ-X^- →Co(III) CT transition and the diagram shows clearly that the σ-bonding is strongly destabilized. Empirically,[2b)] it appears that homolytic cleavage of the Co-Br bond obtains just as homolytic M–M bond cleavage obtains upon $\sigma_b \to \sigma^*$ excitations.

For a long while it was believed that Co(III)-ammines have little LF excited state chemistry and $Co(CN)_5X^{3-}$ species have no CT excited state chemistry. Now several studies[184, 185)] show that $Co(CN)_5X^{3-}$ species do have chemistry associated with $X^- \to$Co(III) CT excitations and usually $Co(CN)_5^{3-}$ results. Despite the fact that Rh(II) is not a common oxidation state, Rh(III) complexes also undergo photoreduction principally by irradiation into ligand→Rh(III) CT absorption bands.[186)] Thus it appears that a fairly large number of nd^6 systems can be reduced to nd^7 systems, but the thermal chemistry following photoreduction will vary depending on the particular complex involved.

Relatively recently[187)] the first photoreduction of a Cr(III) complex was reported. The release of N_3 radicals from $Cr(NH_3)_5N_2^{2+}$, reaction (66) was interpreted

$$Cr(NH_3)_5N_3^{2+} \xrightarrow{h\nu} Cr(NH_3)_5^{2+} + N_3 \qquad (66)$$

as the essential result of irradiation into the $N_3 \to$Cr(III) CT absorption. Like $Co(NH_3)_5^{2+}$, the $Cr(NH_3)_5^{2+}$ is very substitution labile and decomposes to yield the Cr^{2+}_{aq} ion and ammonia. The N_3 radicals yield N_2, and thus the observation of N_2 gas evolution from azide complexes cannot be regarded as evidence for a coordinated nitrene. Photoreduction of $Co(NH_3)_5N_3^{2+}$, $Co(CN)_5N_3^{3-}$, and $Cr(NH_3)_5N_3^{2+}$ all likely occur by similar mechanisms and none of these species gives rise to a coordinated nitrene intermediate.

A large number of other metal complexes have received long and detailed attention, but activity in recent years has revealed few new principles appropriate for discussion here and some systems have been treated in detail elsewhere.[2)] Included among these are oxalato complex photochemistry where oxidation of the oxalato ligands is coupled to the central metal reduction; Ag(I) photochemistry related to imaging systems; uranyl ion photochemical reactions coupled to organic oxidations; and aquo ion photoredox reactions. Two specific topics have recently emerged as potentially important aspects of photoredox behavior in inorganic systems: (1) photoinduced reductive elimination, and (2) photochemistry in wet photoelectric cells. These two topics will be briefly discussed below.

Oxidative addition, such as that exemplified by addition of H-SiR_3 to photogenerated $Fe(CO)_4$, reaction (17), is a very important, large thermal reaction class. Generally, it can be said that oxidative addition to low valent, coordinatively unsaturated metal complexes is common. Recently, Geoffroy, Hammond, and Gray[188a)] have established that *reductive elimination* can be photochemically induced. A typical example is shown in reaction (67). Other small molecules can be reductively eliminated from such Ir(III) complexes to yield the four-coordinate Ir(I) product. Importantly, some photochemically induced reductive eliminations do not occur readily by heating. For example, the photoinduced reductive elimination could be obtained at 77 °K! The mechanism of the reaction is not known, but the authors[188a)] speculated that the Ir(III) complexes have lowest excited states substantially Ir(III)→phosphine CT in character tending to yield an "Ir(IV)" species from which reductive elimination is very fast compared to the rate from the Ir(III) ground state. We note with interest the observation of loss of two methyl radicals by irradiation of $(\eta^5\text{-}C_5H_5)_2M(CH_3)_2$ (M = Ti, Zr, Hf)[188b)], and the apparent photoinduced reductive elimination of H_2 from $(\eta^5\text{-}C_5H_5)_2WH_2$[188c, d)]. The generality of the reductive elimination, the mechanism, and the importance of this reaction in photocatalysis merits serious effort.

$$H_2Ir(diphos)_2^+ \xrightarrow{h\nu} Ir(diphos)_2^+ + H_2 \qquad (67)$$

Lastly, inorganic based photoredox processes on semiconductor electrodes has recently sparked considerable interest in that these can be coupled to important redox

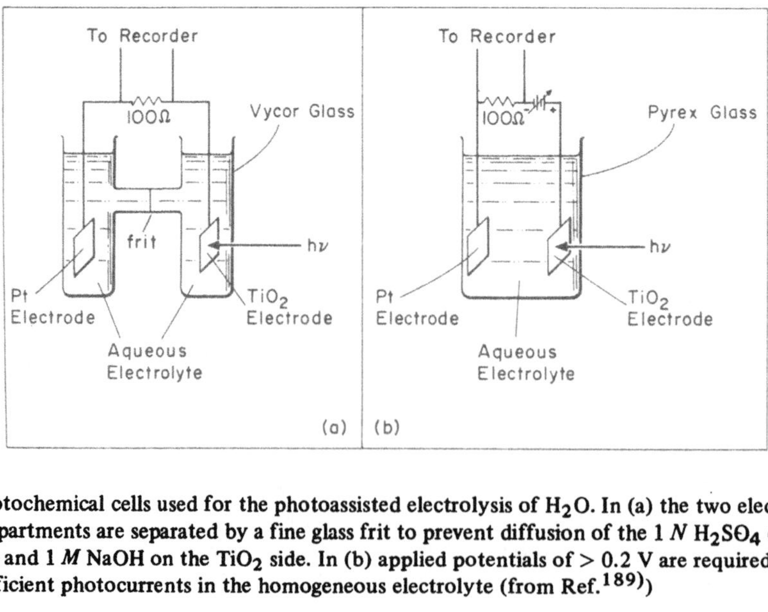

Fig. 9. Photochemical cells used for the photoassisted electrolysis of H_2O. In (a) the two electrode compartments are separated by a fine glass frit to prevent diffusion of the 1 *N* H_2SO_4 on the Pt side and 1 *M* NaOH on the TiO_2 side. In (b) applied potentials of > 0.2 V are required to observe efficient photocurrents in the homogeneous electrolyte (from Ref.[189)])

chemistry. The sustained photoassisted electrolysis of H_2O, reaction (68), has recently been demonstrated[189] by using an electrode system like that schemed in

$$H_2O \xrightarrow[\text{[photoassistance agent]}]{h\nu} H_2 + 1/2\,O_2 \qquad (68)$$

Fig. 9. Upon irradiation of the *n*-type semiconductor TiO_2 with near UV light (>410 nm which corresponds to the band gap energy of TiO_2[190]) O_2 *from the H_2O is evolved at TiO_2* and H_2 is evolved at the Pt electrode which is not irradiated. The source of the O_2 was established by using ^{18}O labelled H_2O, and importantly, no weight loss in the TiO_2 could be detected even when more moles of O_2 were evolved than moles of TiO_2 originally present. Though photoprocesses on semiconductors have been studied for some time in order to characterize the electronic properties of the semiconductor,[191] it has not been until recently[189, 192–194] that attention has turned to energy conversion or chemical synthesis. Preliminary results on the TiO_2 system[192] led the authors to propose that photogeneration of the minority carrier (holes) at the TiO_2 surface lead to the oxidation of H_2O to evolve O_2. The minority carriers in *p*-type semiconductors are electrons. Thus, photolysis at the band gap energy in *p*-type materials should lead to strongly reducing sites on the surface, and photoinduced H_2 production at the photoreceptor could, in principle, be achieved. Until recently, the TiO_2 photoelectrode was the only material which could be used as a photoelectrode to sustain reaction (68), but SnO_2 single crystals have been found[195] to operate in a similar way. Some additional significant work on photoelectrochemical cells has been recently reported[196]. Prospects are bright for useful, photochemically driven redox processes sustained by the photolysis of electrodes in wet electrochemical cells.

Acknowledgement. I thank Professor Harry B. Gray for pointing out the possible role of quintet excited states in complexes, and I thank Professor Edward I. Solomon and Dr. Steven W. Kaiser for reading the final manuscript and offering their suggestions. Work carried out in my laboratories and cited in the references herein has been supported by the National Science Foundation, Office of Naval Research and, National Aeronautics and Space Administration.

V. References

[1] Fellow of the Alfred P. Sloan Foundation, 1974–1976.
[2] Recent books and reviews include: a) Wrighton, M.: Chem. Rev. *74*, 401 (1974);
b) Balzani, V., Carassiti, V.: Photochemistry of coordination compounds. New York: Academic Press 1970;
c) Koerner von Gustorf, E., Grevels, F.-W.: Topics in Current Chemistry *13*, 366 (1969);
d) Balzani, V., Moggi, L., Manfrin, M. F., Bolleta, F., Laurence, G. S.: Coord. Chem. Rev.: in press;
e) Ford, P. C., Petersen, J. D., Hintz, R. E.: Coord. Chem. Rev. *14*, 67 (1974);
f) "Concepts of inorganic photochemistry", A. W. Adamson and P. D. Fleischauer, eds. New York: John Wiley 1975.
[3] a) Ballhausen, C. J.: Introduction to ligand field theory. New York: McGraw-Hill Book Company 1962;
b) Figgis, B. N.: Introduction to ligand fields. New York: Interscience Publishers 1966;
c) Ferguson, J.: Progr. Inorg. Chem. *12*, 159 (1970);
d) Jørgensen, C. K.: ibid. *12*, 101 (1970);
e) Jørgensen, C. K.: Modern aspects of ligand field theory. North Holland Publisher 1971;
f) Orgel, L. E.: An introduction to transition metal chemistry, ligand field theory, 2nd edit. Methuen and John Wiley 1966;
g) Lever, A. B. P.: Inorganic electronic spectroscopy. Elsevier Publishing Co. 1968;
h) Schlafer, H. L., Gliemann, G.: Basic Principles of Ligand Field Theory. New York: Interscience Publishers 1969.
[4] a) Langford, C. H., Gray, H. B.: Ligand substitution processes, New York: W. A. Benjamin 1966;
b) Basolo, F., Pearson, R. G.: Mechanisms of inorganic reactions, 2nd ed. New York: John Wiley and Sons 1967.
[5] Wrighton, M., Gray, H. B., Hammond, G. S.: Mol. Photochem. *5*, 165 (1973).
[6] Wrighton, M., Ginley, D. S., Schroeder, M. A., Morse, D. L.: Pure Appl. Chem.: *41*, 671 (1975).
[7] Adamson, A. W.: J. Phys. Chem. *71*, 798 (1967).
[8] Zink, J. I.: J. Amer. Chem. Soc. *94*, 8039 (1972).
[9] a) Zink, J. I.: Mol. Photochem. *5*, 151 (1973);
b) Zink, J. I.: Inorg. Chem. *12*, 1018 and 1957 (1973);
c) Zink, J. I.: J. Amer. Chem. Soc. *96*, 4464 (1974);
d) Zink, J. I.: Inorg. Chem. *13*, 2489 (1974).
[10] Rösch, N., Messmer, R. O., Johnson, K. H.: J. Amer. Chem. Soc. *96*, 3855 (1974).
[11] a) Perumareddi, J. R.: Coord. Chem. Rev. *4*, 73 (1969);
b) Lever, A. B. P.: ibid. *3*, 141 (1968);
c) Witzke, H.: Theor. Chim. Acta *20*, 171 (1971);
d) Yamatera, H.: Bull. Chem. Soc. (Japan) *31*, 95 (1958);
e) Rowley, D. A.: Inorg. Chem. *10*, 397 (1971);
f) Fleischauer, P. F., Fleischauer, P.: Chem. Rev. *70*, 199 (1970);
g) Forster, L. S., in: Transition metal chemistry, (ed. R. L. Carlin), Vol. 5. New York: Marcel Dekker 1969.
[12] Plane, R. A., Hunt, J. P.: J. Amer. Chem. Soc. *79*, 3343 (1957).
[13] a) Schlafer, H. L.: Z. Physik. Chem. (Frankfurt) *11*, 65 (1957);
b) J. Phys. Chem. *69*, 2201 (1965).
[14] Sabbatini, N., Balzani, V.: J. Amer. Chem. Soc. *94*, 7587 (1972).
[15] Chiang, A., Adamson, A. W.: J. Phys. Chem. *72*, 3827 (1968).
[16] Wilkinson, F.: Advan. Photochem. *3*, 241 (1964).
[17] Wasgestian, H. F.: J. Phys. Chem. *76*, 1947 (1972); cf. also the following for references for similar but less conclusive experiments on other Cr(III) complexes:
a) Kane-Maguire, N. A. P., Langford, C. H.: J. Amer. Chem. Soc. *94*, 2125 (1972);
b) Langford, C. H., Tipping, L.: Can. J. Chem. *50*, 887 (1972);
c) Chen, S., Porter, G. B.: J. Amer. Chem. Soc. *92*, 3196 (1970) and *Chem. Phys. Lett. 6*, 41 (1970);

d) Ballardini, R., Varani, G., Wasgestian, H. F., Moggi, L., Balzani, V.: J. Phys. Chem., *77*, 2947 (1973).
18) Pyke, S. C., Linck, R. G.: J. Amer. Chem. Soc. *93*, 5281 (1971).
19) a) Wasgestian, H. F., Schlafer, H. L.: Z. Phys. Chem. (Frankfurt) *57*, 282 (1968) and *62*, 127 (1968);
b) Ref.[26]), p. 97, 98;
c) Manfrin, M. F., Moggi, L., Balzani, V.: Inorg. Chem. *10*, 207 (1971).
20) Zinato, E., Riccieri, P., Adamson, A. W.: J. Amer. Chem. Soc. *96*, 375 (1974).
21) Ref.[2b]), p. 97, 98.
22) Riccieri, P., Schlafer, H. L.: Inorg. Chem. *9*, 727 (1970)
23) Zinato, E., Furlani, C., Lanna, G., Riccieri, P.: ibid. *11*, 1746 (1972).
24) Zinato, E., Lindholm, R. D., Adamson, A. W.: J. Amer. Chem. Soc. *91*, 1076 (1969).
25) Zinato, E., Riccieri, P.: Inorg. Chem. *12*, 1451 (1973).
26) Kirk, A. D., Moss, K. C., Valentine, J. G.: Can. J. Chem. *49*, 1524 (1971).
27) Kirk, A. D.: Mol. Photochem. *5*, 127 (1973).
28) Kutal, C., Adamson, A. W.: Inorg. Chem. *12*, 1990 (1973).
29) a) Behrendt, S., Langford, C. H., Frankel, L. S.: J. Amer. Chem. Soc. *91*, 2236 (1969);
b) Sastri, V. S., Henwood, R. W., Behrendt, S., Langford, C. H.: ibid. *94*, 753 (1972).
30) Johnson, D. A., Sharpe, A. G.: J. Chem. Soc. A, 798 (1966).
31) Jørgensen, C. K.: Advan. Chem. Phys. *5*, 33 (1963).
32) Forrester, J. S., Ayres, G. H.: J. Phys. Chem. *63*, 1979 (1959).
33) Schmidtke, H. H.: Z. Phys. Chem. (Frankfurt) *40*, 96 (1964).
34) Geoffroy, G. L., Wrighton, M. S., Hammond, G. S., Gray, H. B.: Inorg. Chem. *13*, 430 (1974).
35) Schmidtke, H. H.: J. Mol. Spectroscopy *11*, 483 (1966).
36) Jørgensen, C. K.: Acta Chem. Scand. *10*, 518 (1956).
37) Wheeler, T. E., Perros, T. P., Naeser, C. R.: J. Amer. Chem. Soc. *77*, 3488 (1955).
38) Konig, E., Schlafer, H. L.: Z. Phys. Chem. (Frankfurt) *34*, 355 (1962).
39) Mercer, E. E., Buckley, R. R.: Inorg. Chem. *4*, 1692 (1965).
40) Ref.[3g]), p. 302.
41) Beach, N. A., Gray, H. B.: J. Amer. Chem. Soc. *90*, 5713 (1968).
42) Wrighton, M. S., Handeli, D. I., Morse, D. L.: Inorg. Chem. *15*, (1976).
43) Cf. Jørgensen, C. K.: Absorption spectra and chemical bonding in complexes, pp. 107–109. Oxford: Pergamon Press.
44) Wentworth, R. A. D., Piper, T. S.: Inorg. Chem. *4*, 709 (1965).
45) Cf. Ref.[3a]), pp. 106–108.
46) Ref.[2b]), pp. 231–236.
47) Pribush, R. A., Poon, C. K., Bruce, C. M., Adamson, A. W.: J. Amer. Chem. Soc. *96*, 3027 (1974).
48) Moggi, L., Bolletta, F., Balzani, V., Scandola, F.: J. Inorg. Nucl. Chem. *28*, 2589 (1966).
49) Wrighton, M., Bredesen, D., Hammond, G. S., Gray, H. B.: Chem. Commun., 1018 (1972).
50) Hipps, K. W., Crosby, G. A.: Inorg. Chem. *13*, 1543 (1974).
51) Adamson, A. W., Sporer, A. H.: J. Amer. Chem. Soc. *80*, 3865 (1958), and J. Inorg. Nucl. Chem. *8*, 209 (1958) and see Ref.[48])
52) Adamson, A. W., Chiang, A., Zinato, E.: J. Amer. Chem. Soc. *91*, 5468 (1969).
53) Wrighton, M., Bredesen, D.: Inorg. Chem. *12*, 1707 (1973).
54) Wrighton, M., Gray, H. B., Hammond, G. S., Miskowski, V. M.: ibid. *12*, 740 (1973).
55) a) Miskowski, V. M., Gray, H. B.: ibid. *14*, 401 (1975);
b) Gutterman, D. F., Gray, H. B.: J. Amer. Chem. Soc. *93*, 3364 (1971).
56) a) Viane, L., D'Olieslager, J., De Jaegre, S.: Bull. Soc. Chim. Belg. *82*, 569 1973);
b) idem., manuscript No. 28 at Vth IUPAC Symposium on Photochemistry, Enschede, The Netherlands, July 21–27, 1974, pp. 90–93 of Manuscripts of Contributed Papers.
57) Chen, H.H., Tsao, M-S., Gaver, R.W., Tewari, P.H., Wilmarth, W.K.: Inorg. Chem. *5*, 1913 (1966).
58) a) Kelly, T. L., Endicott, J. F.: J. Amer. Chem. Soc. *94*, 278 (1972);
b) Moggi, L.: Gazz. Chim. Ital. *97*, 1089 (1967).
59) Bauer, R. A., Basolo, F.: J. Amer. Chem. Soc. *90*, 2437 (1968), and Inorg. Chem. *8*, 2231 (1969).

60) Kutal, C., Adamson, A. W.: Inorg. Chem. *12*, 1454 (1973).
61) Muir, M. M., Huang, W.-L.: Inorg. Chem. *12*, 1831 and 1930 (1973).
62) Sheridan, P. S., Adamson, A. W.: J. Amer. Chem. Soc. *96*, 3032 (1974).
63) Sheridan, P. S., Adamson, A. W.: Inorg. Chem. *13*, 2482 (1974).
64) Shirom, M., Stein, G.: J. Chem. Phys. *55*, 3379 (1971).
65) Wrighton, M., Hammond, G. S., Gray, H. B.: J. Amer. Chem. Soc. *93*, 5254 (1971).
66) a) Chaisson, D. A., Hintze, R. E., Stuermer, D. H., Petersen, J. D., McDonald, D. P., Ford, P. C.: ibid. *94*, 6665 (1972);
b) Malouf, G., Ford, P. C.: ibid. *96*, 601 (1974).
67) McClean, R. A. N.: J. Chem. Soc. (Dalton), 1568 (1974).
68) a) Wrighton, M., Hammond, G. S., Gray, H. B.: J. Amer. Chem. Soc. *93*, 4336 (1971);
b) Wrighton, M. S., Morse, D. L., Gray, H. B., Ottesen, D. K.: J. Amer. Chem. Soc.: in press.
69) Saito, H., Fujita, J., Saito, K.: Bull. Chem. Soc. Japan *41*, 359 (1968).
70) Wrighton, M.: Inorg. Chem. *13*, 905 (1974).
71) Wrighton, M. S., Morse, D. L.: J. Organometal. Chem. *97*, 405 (1975).
72) a) Chatt, J., Gamlen, G. A., Orgel, L. E.: J. Chem. Soc., 486 (1958);
b) Gray, H. B., Ballhausen, C. J.: J. Amer. Chem. Soc. *85* 260 (1963);
c) Gray, H. B., Progr. Transition Metal Chem. *1*, 240 (1965);
d) Martin, Jr., D. S., Tucker, M. A., Kassman, A. J., Inorg. Chem. *4*, 1682 (1965) and *5*, 1298 (1966);
e) McCaffery, A. J., Schatz, P. N., Stephens, P. J.: J. Amer. Chem. Soc. *90*, 5730 (1968);
f) Mason, W. R., Gray, H. B.: ibid. *90*, 5721 (1968);
g) Piepho, S. B., Schatz, P. N., McCaffery, A. J.: ibid. *91*, 5994 (1969);
h) Preer, J. R., Gray, H. B.: ibid. *92*, 7306 (1970);
i) Dawson, J. W., Venanzi, L. M., Preer, J. R., Hix, J. E., Gray, H. B.: ibid. *93*, 778 (1971);
j) Dawson, J. W., Gray, H. B., Hix, Jr., J. E., Preer, J. R., Venanzi, L. M.: ibid. *94*, 2979 (1972);
k) Geoffroy, G. L., Wrighton, M. S., Hammond, G. S., Gray, H. B.: ibid. *96*, 3105 (1974);
l) Furlani, C.: Coord. Chem. Rev. *3*, 141 (1968);
m) Sacconi, L.: Pure Appl. Chem. *17*, 95 (1968);
n) Ciampolini, M.: Struct. Bonding *6*, 52 (1969).
73) Scandola, F., Traverso, O., Carassiti, V.: Mol. Photochem. *1*, 11 (1969).
74) Natarjan, P., Adamson, A. W.: J. Amer. Chem. Soc. *93*, 5599 (1971).
75) Bartocci, C., Scandola, F., Balzani, V.: ibid. *91*, 6948 (1969).
76) a) Poliakoff, M., Turner, J. J.: J. Chem. Soc. (Dalton), 2276 (1974);
b) Stolz, I. W., Dobson, G. R., Sheline, R. K.: J. Amer. Chem. Soc. *84*, 3589 (1962) and *85*, 1013 (1963).
77) Moss, J. R., Graham, W. A. G.: Chem. Commun., 835 (1970).
78) Fischer, E. O., Bittler, K.: Naturforsch B *30*, 254 (1975).
79) a) Rest, A. J.: Chem. Commun., 345 (1970);
b) Keeton, D. P., Basolo, F.: Inorg. Chim. Acta *6*, 33 (1972).
80) Udovich, C. A., Clark, R. J.: Inorg. Chem. *8*, 938 (1969).
81) Dartiguenave, M., Dartiguenave, Y., Gray, H. B.: Bull. Chim. Soc. Fr. *12*, 4223 (1969).
82) Schubert, E. H., Sheline, R. K.: Inorg. Chem. *5*, 1071 (1966).
83) Organic synthesis via metal carbonyls, (eds. I. Wender and P. Pino), Vol. 1. New York: Interscience Publishers 1968.
84) Wrighton, M. S., Haverty, J. L.: Z. Naturforsch. B.: *30*, 254 (1975).
85) Jetz, W., Graham, W. A. G.: Inorg. Chem. *10*, 4 (1971).
86) Valentine, D., Valentine, J., Sarver, B. in: Photoexcited states, (ed. J. N. Pitts). New York: Gordon and Breach 1970.
87) Oliver, A. J., Graham, W. A. G.: Inorg. Chem. *10*, 1 (1971).
88) Massey, A. G., Orgel, L. E.: Nature *191*, 1387 (1961).
89) Wrighton, M. S., Ellis, A. B.: unpublished results.
90) a) Terenin, A. N: Propl. Kinet. Katal. Akad. Nawk., USSR *8*, 17 (1955);
b) Terenin, A. N., Solonitzin, Y.: Disc. Faraday Soc. *28*, 28 (1959);
c) Lange, W. J.: J. Vacuum Sci, Tech. *2*, 74 (1965);

d) Adams, R. O., Donaldson, E. E.: J. Chem. Phys. *42*, 770 (1965).
91) a) Modell, M., Baddour, R. F.: J. Phys. Chem. *74*, 1392 (1970);
b) Modell, M., Ph. D.: Thesis, Massachusetts Institute of Technology 1964;
c) Aleksandrowicz, J., Ph. D.: Thesis, Massachusetts Institute of Technology 1970.
92) a) Cotton, F. A.: Rev. Pure Appl. Chem. *17*, 25 (1967); Accounts Chem. Res. *2*, 240 (1969); Chem. Soc. Rev.: in press;
b) Penfold, B. R.: Perspectives in structural chemistry, (eds. J. Dunitz and J. A. Ibers), Vol. 2, p. 68. New York: John Wiley & Sons 1968;
c) Baird, M. C.: Progr. Inorg. Chem. *9*, 1 (1968);
d) Kepert, D. L., Vrieze, K. in: Halogen chemistry, (ed. V. Guttman), Vol. 3, p. 1. New York: Academic Press 1967;
e) Johnston, R. D.: Adv. Inorg. Chem. Radiochem. *13*, 471 (1970);
f) Chini, P.: Inorg. Chim. Acta Rev. *2*, 31 (1968);
g) Abel, E. W., Stone, F. G. A.: Quart. Rev. *23*, 325 (1969);
h) Walton, R. A.: Progr. Inorg. Chem. *16*, 9 (1973).
93) Bidinosti, D. R., McIntyre, N. S.: Chem. Comm., 1 (1967).
94) Cotton, F. A., Harris, C. B.: Inorg. Chem. *6*, 924 (1967).
95) Cf. Cotton, F. A., Wilkinson, G.: Advanced inorganic chemistry, 3rd edit., pp 550–553. New York: Interscience Publishers 1972.
96) a) Cotton, F. A., Curtis, N. F., Robinson, W. R.: Inorg. Chem. *4*, 1969 (1965);
b) Cotton, F. A., Robinson, W. R., Walton, R. A., Whyman, R.: ibid. *6*, 929 (1967);
c) Cotton, F. A., Oldham, C., Walton, R. A.: ibid. *6*, 214 (1967);
d) van Bronswyk, W.: Chem. Soc. A, 692 (1968);
e) DeWit, D., Fawcett, J. P., Poe, A. J., Twigg, M. W.: Coord. Chem. Rev. *8*, 81 (1972).
97) Calvert, J. G., Pitts, J. N: Photochemistry. New York: John Wiley & Sons 1966.
98) Wrighton, M. S., Ginley, D. S.: J. Amer. Chem. Soc. *97*, 2065 (1975).
99) Bamford, C. H., Paprotny, J.: Chem. Comm., 140 (1971).
100) Wrighton, M., Bredesen, D.: J. Organometal. Chem. *50*, C35 (1973).
101) Evans, G. O., Sheline, R. K.: J. Inorg. Nucl. Chem., 2862 (1968).
102) Wrighton, M. S., Ginley, D. S.: J. Amer. Chem. Soc., *97*, 4246 (1975).
103) Ginley, D. S., Wrighton, M. S.: ibid. *97*, 4908 (1975).
104) King, R. B., Bisnette, M. B.: Inorg. Chem. *3*, 785 (1964).
105) Cotton, F. A., Deeming, A. J., Josty, P. L., Ullah, S. S., Domingos, A. J. P., Johnson, B. F. G., Lewis, J.: J. Amer. Chem. Soc. *93*, 4621 (1971).
106) a) Knox, S. A. R., Stone, F. G. A.: J. Chem. Soc., A, 2874 (1971); 3147 (1970) and 2559 (1969);
b) Brockes, A., Knox, S. A. R., Stone, F. G. A., ibid., 3469 (1971).
107) Johnson, B. F. G., Lewis, J., Twigg, M. V.: J. Organometal. Chem. *67*, C75 (1974).
108) Wrighton, M. S., Ginley, D. S.: unpublished results.
109) a) Levenson, R. A., Gray, H. B., Caesar, G. P.: J. Amer. Chem. Soc. *92*, 3653 (1970);
b) Levenson, R. A., Ph. D.: Thesis, Columbia University 1970;
c) Levenson, R. A., Gray, H. B.: J. Amer. Chem. Soc.: *97*, 6042 (1975).
110) Hieber, W., Schrop, Jr., W.: Z. Naturforsch. B *15*, 271 (1960).
111) a) Halpern, J., Maher, J. P.: J. Amer. Chem. Soc. *87*, 5361 (1965);
b) Chock, P. B., Halpern, J.: ibid. *91*, 582 (1969);
c) Kwiatek, J., Seyler, J. K.: J. Organometal. Chem. *3*, 421 (1965).
112) Hallock, S. A., Wojcicki, A.: J. Organometal. Chem. *54*, C27 (1973).
113) Huber, H., Kundig, E. P., Ozin, G. A.: J. Amer. Chem. Soc. *96*, 5585 (1974).
114) Fischer, E. O., Offhaus, E., Muller, J., Nothe, D.: Chem. Ber. *105*, 1027 (1972).
115) Fieldhouse, S. A., Fullam, B. W., Neilson, G. W., Symons, M. C. R., J. Chem. Soc. (Dalton), 567 (1974).
116) a) Osborne, A. G., Stiddard, M. H. B.: J. Chem. Soc., 634 (1964);
b) Miller, J. R., Myers, D. H.: Inorg. Chim. Acta *5*, 215 (1970);
c) Lewis, J., Nyholm, R. S., Osborne, A. G., Sandhu, S. S., Stiddard, M. H. B.: Can. J. Chem. *45*, 161 (1967);

e) Clark, R. J., Hargaden, J. P., Haas, H., Sheline, R. K.: Inorg. Chem. *7*, 673 (1968);
f) Bor, G.: Chem. Comm., 641 (1965);
g) Ziegler, M. L., Haas, H., Sheline, R. K.: Chem. Ber. *98*, 2454 (1965);
h) Crow, J. P., Cullen, W. R., Hou, F. L.: Inorg. Chem. *11*, 2125 (1972);
i) Crow, J. P., Cullen, W. R., Hou, F. L., Chan, L. Y. Y., Einstein, F. W. B.: Chem. Comm., 1229 (1971).

117) a) King, R. B., Pannell, K. H.: Inorg. Chem. *7*, 2356 (1968);
b) Barnett, K. W., Treichel, P. M.: ibid. *6*, 294 (1967).

118) Haines, R. J., Nyholm, R. S., Stiddard, M. G. B.: J. Chem. Soc. A, 43 (1968).

119) Burkett, A. R., Meyer, T. J., Whitten, D. G.: J. Organometal. Chem. *67*, 67 (1974); cf. however, Ref.[102] and Hughey, J. L., IV, Bock, C. R., Meyer, T. J.: J. Amer. Chem. Soc. *97*, 4440 (1975).

120) Geoffroy, G. L., Hammond, G. S., Gray, H. B.: J. Amer. Chem. Soc. *96*, 5565 (1974).

121) a) Cotton, F. A., *et al.*: Science *145*, 1035 (1964);
b) Cotton, F. A.: Inorg. Chem. *4*, 334 (1965).

122) Cowman, C. D., Gray, H. B.: J. Amer. Chem. Soc. *95*, 8177 (1973).

123) Wrighton, M. S., Morse, D. L., Pdungsap, L.: ibid. 2073 (1975).

124) a) Crosby, G. A., Carstens, D. H. W.: Molecular luminescence, ed. E. L. Linn. New York: W. A. Benjamin Co. 1969;
b) DeArmond, M. K., Hillis, J. E., J. Chem. Phys. *49*, 466 (1968).

125) a) Whitten, D. G., McCall, M. T.: ibid. *91*, 5097 (1969);
b) Bartocci, G., Bortolus, P., Mazzucato, U.: J. Phys. Chem. *77*, 607 (1973);
c) Favaro, G., Mazzucato, U., Masetti, F.: ibid. *77*, 601 (1973).

126) a) Yamada, Y., Milijkovic, D., Wehrli, P., Golding, B., Loliger, P., Keese, R., Muller, K., Eschenmoser, A.: Angew. Chem. Int. Ed. *8*, 343 (1969);
b) Eschenmoser, A.: Pure Appl. Chem. *20*, 1 (1969) and Quart. Rev. *24*, 366 (1970);
c) Gotschi, E., Eschenmoser, A.: Angew. Chem. Int. Ed. *12*, 912 (1973);
d) Woodward, R. B.: Pure App. Chem. *33*, 145 (1970);
e) Eschenmoser, A.: Pure Appl. Chem. Suppl. XXIII IUPAC Congress, Boston, Vol. 3, 69 (1971).

127) a) Salomon, R. G., Folting, K., Streib, W. E., Kochi, J. K.: J. Am. Chem. Soc. *96*, 1145 (1974).
b) Salomon, R. G., Kochi, J. K.: ibid. *96*, 1137 (1974).

128) a) Zarnegar, R. R., Bock, C. R., Whitten, D. G.: ibid. *95*, 4367 (1973);
b) Zarnegar, P., Whitten, D. G.: ibid. *93*, 3776 (1971).

129) Wrighton, M., Hammond, G. S., Gray, H. B.: Mol. Photochem. *5*, 179 (1973).

130) Wrighton, M. S. Pdungsap, L.: unpublished results.

131) Finnegan, R. A., Mattice, J. J.: Tetrahedron *21*, 1015 (1965).

132) Bozak, R. E.: Chem. Ind. (London), 24 (1969).

133) a) Rausch, M., Vogel, M., Rosenberg, H.: J. Org. Chem. *22*, 903 (1957);
b) Weliky, N., Gould, E. S.: J. Am. Chem. Soc. *79*, 2742 (1957).

134) Richards, J. H., Pisker-Trifunac, N.: Proc. Paint Res. Inst. *41*, 363 (1969).

135) Ref.[2b], p. 349.

136) Wrighton, M., Hammond, G. S., Gray, H. B.: J. Am. Chem. Soc. *92*, 6068 (1970) and *93*, 3285 (1971).

137) Wrighton, M., Hammond, G. S., Gray, H. B.: J. Organometal. Chem. *70*, 283 (1974).

138) a) Hill, B., Math, K., Pillsbury, D., Voecks, G., Jennings, W.: Mol. Photochem. *5*, 195 (1973);
b) Jennings W. and Hill, B., J. Am. Chem. Soc. *92*, 3199 (1970).

139) Wrighton, M. S., Schroeder, M. A.: J. Am. Chem. Soc. *96*, 6235 (1974).

140) a) Wrighton, M., Schroeder, M. A.: J. Am. Chem. Soc. *95*, 5764 (1973);
b) Nasielski, J., Kirsch, P., Wilputte-Steinert, L.: J. Organometal. Chem. *27*, C13 (1971);
c) Platbrood, G., Wilputte-Steinert, L.: J. Organometal. Chem. *70* 393 and 407 (1974).

141) Schenck, G. O., Koerner von Gustorf, E., Junn, M. J.: Tetrahedron Lett., 1059 (1962).

142) Tsay, F. D., Gray, H. B., Danon, J.: J. Chem. Phys. *54*, 3760 (1971).

143) a) Halpern, J.: Accounts Chem. Res. *3*, 386 (1970);
b) Griffith, W. P., Wilkinson, G.: J. Chem. Soc., 1629 (1959);

c) Kimball, M. E., Martella, J. P., Kaska, W. C.: Inorg. Chem. *6*, 414 (1967).
144) a) Asinger, F., Fell, B., Schrage, K.: Chem. Ber. *98*, 381 (1965) and *98*, 372 (1965);
b) Asinger, F., Fell, B., Collin, G.: ibid. *96*, 716 (1963);
c) Carr, M. D., Kane, V. V., Whiting, M. C.: Proc. Chem. Soc. (London), 408 (1964);
d) Fell, B., Krings, P., Asinger, F.: Chem. Ber. *99*, 3688 (1966);
e) Damico, R.: J. Org. Chem. *33*, 1550 (1968);
f) Jolly, P. W., Stone, F. G. A., Mackenzie, K.: J. Chem. Soc., 6416 (1965);
g) Damico, R., Logan, T. J.: J. Org. Chem. *32*, 2356 (1967);
h) Heimsbach, P., Traunmuller, R.: Angew. Chem. *78*, 604 (1966).
145) Schroeder, M. A., Wrighton, M. S.: J. Amer. Chem. Soc., *97* (1975).
146) a) Frankel, E. N.: J. Org. Chem. *37*, 1549 (1972);
b) Frankel, E. N., Butterfield, R. O.: ibid. *34*, 3930 (1969);
c) Frankel, E. N., Little, F.: J. Amer. Oil Chem. Soc. *46*, 256 (1969);
d) Cais, M., Frankel, E. N., Rejoan, R. A.: Tetrahedron Lett., 1919 (1968).
147) Natta, G., Ercoli, R., Calderazzo, F., Santambrozio, E.: Chim. Imd. (Milan) *40*, 1003 (1958).
148) Schroeder, M. A., Wrighton, M. S.: J. Organometal. Chem. *74*, C29 (1974).
149) Schroeder, M. A., Wrighton, M. S.: unpublished results.
150) Wrighton, M. S., Haverty, J. L.: Z. Naturforsch B.: *30*, 254 (1975).
151) Gafney, H. D., Reed. J. L., Basolo, F.: J. Am. Chem. Soc. *95*, 7998 (1973).
152) Reed, J. L., Gafney, H. D., Basolo, F.: ibid. *96*, 1363 (1974).
153) a) Endicott, J. F., Hoffman, M. Z.: J. Am. Chem. Soc. *90*, 4740 (1968);
b) Adamson, A. W.: Discus. Faraday Soc. *29*, 163 (1960);
c) Ferraudi, G., Endicott, J. F.: Inorg. Chem. *12*, 2389 (1973).
154) Zink, J. I.: Inorg. Chem. *13*, 446 (1975).
155) McPhail, A. T., Knox, G. R., Robertson, C. G., Sim, G. A.: J. Chem. Soc. A, 205 (1971).
156) a) Lytle, F. E., Hercules, D. M.: J. Am. Chem. Soc. *91*, 253 (1969);
b) Veening, H., Brandt, W. W.: Anal. Chem. *32*, 1426 (1960).
157) a) Barker, S. L.: Chem. Commun., 363 (1971);
b) Geiss, W., Schlafer, H. L.: Z. Phys. Chem. (Frankfurt) *65*, 107 (1969);
c) Balzani, V., Ballardini, R., Gandolfi, M. T., Moggi, L.: J. Am. Chem. Soc. *93*, 339 (1971).
158) Wrighton, M., Morse, D. L.: J. Am. Chem. Soc. *96*, 998 (1974).
159) Wunschel, Jr., K. R., Ohnesorge, W. E.: ibid. *89*, 2777 (1967).
160) Crosby, G. A., Hipps, K. W., Elfring, Jr., W. H.: ibid. *96*, 629 (1974).
161) Harrigan, R. W., Crosby, G. A.: J. Chem. Phys. *59*, 3468 (1973).
162) a) Wrighton, M., Markham, J.: J. Phys. Chem. *77*, 3042 (1973);
b) Demas, J. N., Diemente, D., Harris, E. W.: J. Am. Chem. Soc. *95*, 6864 (1973).
163) Demas, J. N., Adamson, A. W.: J. Am. Chem. Soc. *93*, 1800 (1971).
164) Gafney, H. D., Adamson, A. W.: ibid. *94*, 8238 (1972).
165) Bock, C. R., Meyer, T. J., Whitten, D. G.: ibid. *96*, 4710 (1974).
166) Navon, G., Sutin, N.: Inorg. Chem. *13*, 2159 (1974).
167) Laurence, G. S., Balzani, V.: ibid. *13*, 2976 (1974).
168) Demas, J. N., Adamson, A. W.: J. Amer. Chem. Soc. *95*, 5159 (1973).
169) a) Martin, J. E., Hart, E. J., Adamson, A. W., Gafney, H., Halpern, J.: J. Am. Chem. Soc. *94*, 9238 (1972);
b) Hercules, D. M., Lytle, F. E.: ibid. *88*, 4745 (1966);
c) Lytle, F. E., Hercules, D. M.: Photochem. Photobiol. *13*, 123 (1971);
d) Tokel, N. E., Bard, A. J.: J. Am. Chem. Soc. *94*, 2862 (1972).
170) Matheson, M., Mulac, W. A., Rabani, J.: J. Phys. Chem. *67*, 2613 (1963).
171) Shiron, M., Stein, G.: Nature *204*, 778 (1964).
172) a) Ohno, S., Tsushihashi, G.: Bull. Chem. Soc. Japan *38*, 1052 (1965);
b) Ohno, S.: ibid. *40*, 1765, 1770, 1776, 1779 (1967).
173) Airey, P. D., Dainton, F. S.: Proc. Roy. Soc. (London) *A 291*, 340, 478 (1966).
174) Ottolenghi, M., Rabani, J., Stein, G.: Israel J. Chem. *5*, 309 (1967).
175) Devonshire, R., Weiss, J. J.: J. Phys. Chem. *72*, 3815 (1968).
176) Waltz, W. L., Adamson, A. W.: ibid. *73*, 4250 (1969).
177) Shirom, M., Stein, G.: J. Chem. Phys. *55*, 3372 (1971).

178) Emschwiller, G., Legros, J.: Compt. Rend. *261*, 1535 (1965).
179) a) Goodenow, E. L., Garner, C. S.: J. Am. Chem. Soc. *77*, 5268 and 5272 (1955).
180) a) Carassiti, V., Balzani, V.: Ann. Chim Rome *51*, 518 (1961);
b) Balzani, V., Carassiti, V.: ibid. *51*, 533 (1961) and *52*, 432 (1962).
181) a) Ref.[2b], pp. 159–161, 174–178, 315–318;
b) Heidt, L. J., McMillan, A. F.: Science *117*, 75 (1953);
c) Marcus, R. J.: ibid. *123*, 399 (1956).
182) a) Endicott, J. F. in: Concepts of inorganic photochemistry, (eds. A. W. Adamson and P. F. Fleischauer), New York: John Wiley Sons 1975.
b) Endicott, J. F.: Inorg. Chem. *14*, 448 (1975);
c) Ferraudi, G., Endicott, J. F.: J. Am. Chem. Soc. *96*, 3681 (1974).
183) Scandola, F., Bartocci, C., Scandola, M. A.: J. Phys. Chem. *78*, 572 (1974) and J. Am. Chem. Soc. *95*, 7898 (1973).
184) Viaene, L., D'Olieslager, J., DeJaegere, S.: Bull. Soc. Chim. Belg. *82*, 569 (1973).
185) Ferraudi, G., Endicott, J. F.: J. Am. Chem. Soc. *95*, 2371 (1973).
186) Kelly, T., Endicott, J. F.: ibid. *92*, 5733 (1970) and *94*, 1797 (1972).
187) Vogler, A.: ibid. *93*, 5912 (1973).
188) a) Geoffroy, G. L., Hammond, G. S., Gray, H. B.: J. Amer. Chem. Soc. *97*, 3933 (1975);
b) Alt, H., Rausch, M. D.: ibid. *96*, 5936 (1974);
c) Elmitt, K., *et al.*: Chem. Commun. *1974*, 747;
d) Farrugia, L., Green, M. L. H.: ibid. *1975*, 416.
189) Wrighton, M. S., Ginley, D. S., Wolczanski, P. T., Ellis, A. B., Morse, D. L., Linz, A.: Proc. Nat. Acad. Sci., U.S. *72*, 1518 (1975).
190) Cronemeyer, D. C.: Phys. Rev. *87*, 876 (1952).
191) a) Gerischer, H.: Surf. Sci. *13*, 265 (1969);
b) Freund, T., Gomes, W. P.: Catl. Rev. *3*, 1 (1969);
c) Brattain, W. H., Garrett, C. G. B.: Bell Syst. Techn. J. *34*, 129 (1955).
192) Fujishima, A., Honda, K.: Nature *238*, 37 (1972) and Bull. Chem. Soc. Japan *44*, 1148 (1971).
193) Gerischer, H.: J. Electroanal. Chem. *58*, 263 (1975) and references therein.
194) Keeney, J., Weinstein, D. H., Haas, G. M.: Nature *253*, 719 (1975).
195) Wrighton, M. S., Morse, D. L., Ellis, A. B., Ginley, D. S., Abrahamson, H. B.: J. Amer. Chem. Soc. *97* (1975).
196) a) The most efficient photoassisted electrolysis of H_2O using a $SrTiO_3$ photoelectrode: Wrighton, M. S., Ellis, A. B., Wolczanski, P. T., Morse, D. L., Abrahamson, H. B., Ginley, D. S.: ibid *98* (1976);
b) TiO_2 thin-film photoelectrodes: Hardee, K. L., Bard, A. J.: J. Electrochem. Soc. *122*, 739 (1975); Gissler, W., Lensi, P. L., Pizzini, S.: J. Appl. Electrochem. *5* (1975);
c) TiO_2 current-voltage curves: Nozik, A. J.: Nature *257*, 383 (1975);
d) TiO_2 quantum yields: Ohnishi, T., Nakato, Y., Tsubomura, H.: Ber. Bunsenges. Phys. Chem. *79*, 523 (1975);
e) Semiconductors coated with this metal films: Nakato, Y., Ohnishi, T., Tsubomura, H.: Chem. Lett. *1975*, 883.

Received April 7, 1975

Complexation and Activation of Diazenes and Diazo Compounds by Transition Metals

Dr. Angelo Albini* and Dr. Horst Kisch

Institut für Strahlenchemie im Max-Planck-Institut für Kohlenforschung, D-4330 Mülheim a. d. Ruhr

Contents

* Present address: Ist. di Chimica Organica, Università, I-27100 Pavia, Italy.

I. Introduction

Reactions involving the olefinic bond constitute one of the most important and extensively studied fields of organometallic chemistry[1]. In contrast, the reactivity of the isoelectronic diazenes (–N=N–) towards transition metals has until recently been largely unexplored. Much of this current interest in the coordinating properties of diazenes[2] stems from the probable intermediacy of diazene complexes in biological nitrogen-fixation[3] as well as from the nonenzymatic conversion of coordinated dinitrogen into diazene derivatives[4]. This article is mainly concerned with the interaction of metal carbonyls with cyclic diazenes and diazo compounds, and with reactions of the coordinated N=N bond[5]. The closely related diazenato complexes have been recently reviewed[2d].

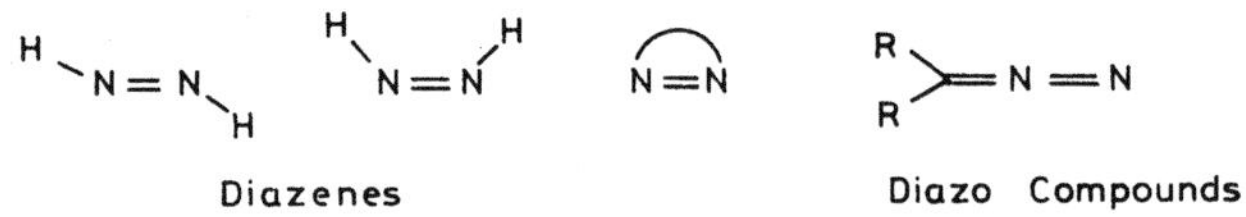

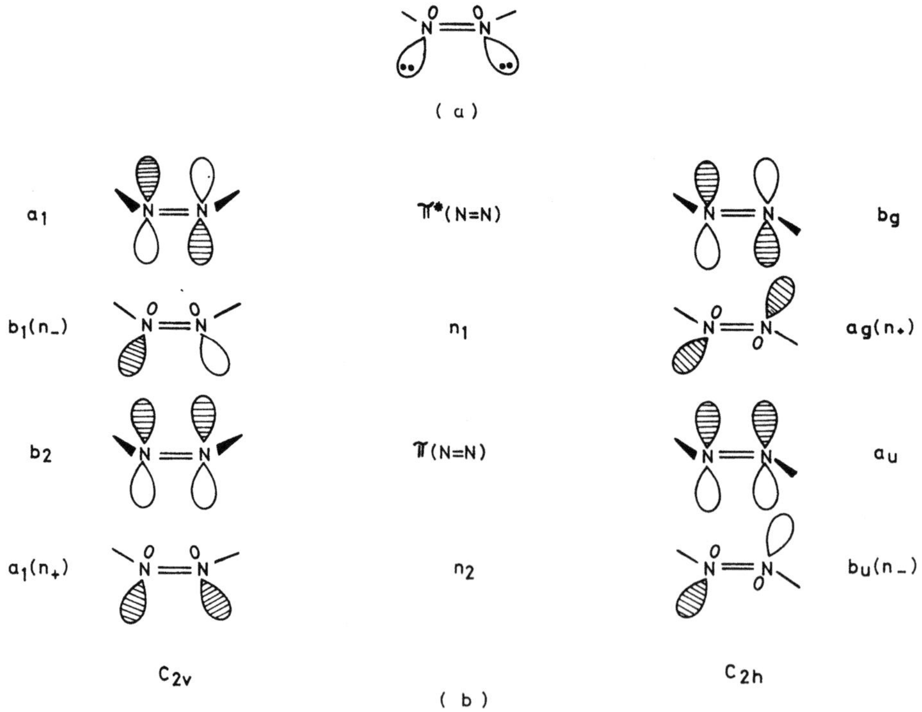

Fig. 1. Classical (a) and schematic MO-description (b) of the cis (C_{2v}) and trans (C_{2h}) diazene group. The symbols in the center of the figure are used in the text. (See also Table 2.)

II. Electronic Structure of Ligands

The basic electronic structure of diazenes has been thoroughly elucidated. In the classical formulation of the $-\ddot{N}=\ddot{N}-$ group the two electron lone pairs are assumed to occupy sp^2 hybridized atomic orbitals which are localized on the nitrogen atoms (*e.g.* Fig. 1a for a cis-diazene). This model implies two energetically degenerated highest occupied orbitals and it has been assumed that the overlap integral between these sp^2 atomic orbitals may be neglected[6]. However, it is generally accepted now that this integral is not neglectable and that interaction of the lone pairs yields a symmetric (n_+) and antisymmetric (n_-) combination[7, 8]. A very high energetical separation of 7 eV has been calculated for trans-diazene (trans-diimine) by ab initio molecular orbital methods[9a, b] (Experimental value[9c]: 5.1 eV). Haselbach and Heilbronner have measured the photoelectron spectra of a variety of diazenes and assignments are based on correlation with spectra of corresponding C=C and C=N-analogues as well as on MO-calculations[10]. From their results it follows that the energetical separation between these nonbonding orbitals is a function of the NNR bond angle, but in most diazenes the π(N=N) bonding orbital lies energetically between n_+ and n_-. The relevant orbitals of an N=N bond are schematically depicted in Fig. 1. Thus, in trans-diazenes (C_{2h} symmetry) the highest occupied MO (HOMO) is the n_+, combination, transforming as a_g, while in *cis*-diazenes (C_{2v}) the antisymmetric n_- combination (b_1) is higher than $n_+(a_1)$. To simplify the notation for these MO's, regardless which symmetry is present, we have chosen to use n_1 as the designation for the energetically higher orbital and n_2 for the lower one[11]. In this designation the n_1 orbital lies always at higher energy whether the compound has cis- or trans-configuration. The splitting between n_1 and n_2 is in the range of 3–3.6 eV for 3- and 5-membered cyclic diazenes[10]. The 4-membered 1.2-diazetines exhibit a very small separation of ~ 1.5 eV and in this case the n_2 orbital lies above the π(N=N) level[10]. These experimental findings confirmed the predictions made by Gimarc[13] who found that, in addition to the "throughspace" interaction[14] present in trans-diazenes, a "throughbond" interaction of the n_2 orbital with a lower lying N—N σ-orbital is possible for cis-diazenes; this interaction raises the energy of the n_2 level and becomes maximal for NNR angles of ~90° as present in the 4-membered 1,2-diazenes[10]. Table 1 contains ionisation potentials and $n_1 \rightarrow \pi^*$ transition energies of some selected diazenes. In agreement with these results the lowest-energy band in the electronic spectrum of diazenes originates from an $n_1 \rightarrow \pi^*$ transition. While there is a general agreement about this assignment, the higher absorption bands are not as securely assigned. In the case of trans-dimethyldiazene the band at 54 kK was first assigned to an $n_2 \rightarrow \pi^*$ transition[20]. However, later experiments showed that the vibrational structure of this band is washed out by nitrogen gas of high pressure (119 atm), and that there is a general shift of intensity to higher frequencies, as expected for an upper Rydberg state. Accordingly, this band was assigned to an $n_1 \rightarrow 3p$ Rydberg transition[21].

Molecular orbital theory not only predicted the interaction between the "lone pairs" but also showed that these are more or less delocalized over the neighbouring bonds[22, 23]. This effect is relatively small for n_1 in trans-dimethyldiazene[23] but

Table 1. Vertical ionisation potentials and $n_1 \rightarrow \pi^*$ transition energies (eV) of diazenes[1])

	$I_v(n_1)$	$I_v(\pi)$	$I_v(n_2)$	$n_1 \rightarrow \pi^*$ [3])
trans-Diazene[2])	10.2	14.39	15.03	
trans-Dimethyldiazene	8.98	11.84	12.3	3.51[4])
trans-bis(isopropyl)-diazene	8.47	~11.1	11.5	3.45[4])
cis-bis(isopropyl)diazene	8.24	11.18	11.9	3.25[4])
Pyridazine[5])	9.31	10.6, 10.9	11.3	3.3[6])
cis-Azobenzene[7])	8.46	8.77	–	2.82
Benzo[c]cinnoline[8])	9.1	10.2, 10.8	11.3	3.1
2,3-Diazanorbornene	8.96	11.53	11.91	3.63
3,3,5,5-Tetramethyl-1-pyrazoline	8.63	10.91	11.26	3.54
3,4-Diaza-exo-bicyclo [4.2.1.0^{2,5}]non-3-ene	8.9	11.3	10.4	–
3,3-Dimethyldiazirine	9.76	12.1	13.3	3.57[9])

[1]) Collected in Ref.[12]).
[2]) Ref.[9c]).
[3]) In *n*-hexane solution.
[4]) In gas phase.
[5]) Ref.[15]).
[6]) Ref.[16a]).
[7]) Ref.[16b]).
[8]) Preliminary values of Refs.[17, 18]).
[9]) Ref.[19]).

becomes larger in cis-dimethyldiazene and diazirine[22]). Table 2 contains the bond characters of the frontier orbitals of these diazenes as obtained by CNDO/2 calculations[24]). These show that in the case of trans-dimethyldiazene the "lone-pairs" contribute about 80% to the HOMO (n_1) which is localized by about 20% on the C–N bonds. This "lone pair character" of the n_1 orbital is only 67% in cis-dimethyldiazene and significantly drops to 32% in diazirine. In the latter case the symmetry allowed mixing of the n_1 linear combination with a C-2*p* orbital is so large that this molecular orbital is to be viewed rather as a bonding σ(C–N) than as a nitrogen lone pair orbital (Fig. 2b). It may be further noted that the lone pair character of the lower lying n_2 orbital is comparable with n_1 in the case of cis-dimethyldiazene, only half (42%) of the n_1-value for trans-dimethyldiazene, but more than two times larger (78%) than n_1 in the case of diazirine. In all diazenes calculated the LUMO (π^*(N=N)) is found to be localized almost completely on the N=N bond (96–100%). Considering these numbers the concept of "lone pairs" largely looses its significance and it seems important to remember these characteristic properties of the frontier orbitals of the N=N bond when discussing its chemical reactivity.

In contrast to diazenes, the electronic structure of diazocompounds was not so intensively studied. The HOMO of diazomethane was calculated to be a $b_1(\pi)$ orbital corresponding to the non-bonding orbital of a π-allyl system[22, 26–28] (Fig. 2). This orbital is responsible for the 1.3-dipolar reactivity of diazoalkanes. Typical ionisation energies are 9.0, 7.88 and 8.33 eV for diazomethane, 2-diazopropane and diazocyclopentadiene, respectively[28].

Table 2. Composition of frontier molecular orbitals in diazenes[24]

	MO	Bond Characters %			
		N=N	N–C	C–H	LP¹)
trans-Dimethyldiazene	$a_g(n_1)$	0	19	2	79
	$a_u(\pi)$	77	0	23	0
	$b_u(n_2)$	0	5	54	42
cis-Dimethyldiazene	$b_1(n_1)$	0	29	4	67
	$b_2(\pi)$	85	0	15	0
	$a_1(n_2)$	9	0	22	69
Diazirine	$b_1(n_1)$	0	68	0	32
	$b_2(\pi)$	74	0	26	0
	$a_1(n_2)$	16	6	0	78

¹) LP = lone pair character.

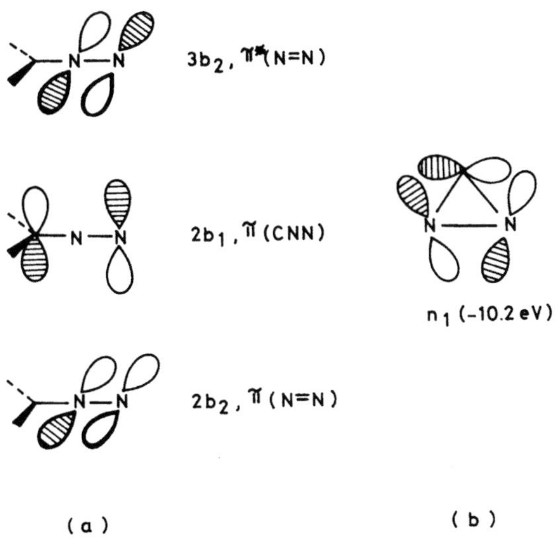

Fig. 2. Schematic drawing of (a) frontier MO's of diazomethane[26,28], (b) HOMO of diazirine[22, 25]

III. Modes of Coordination

From the fact that all diazenes discussed in Section II possess a HOMO(n_1, −8 to −10 eV) which has more or less pronounced lone pair character, it follows that the diazene group should be a good σ-donor forming easily complexes of type I, (Fig. 3).

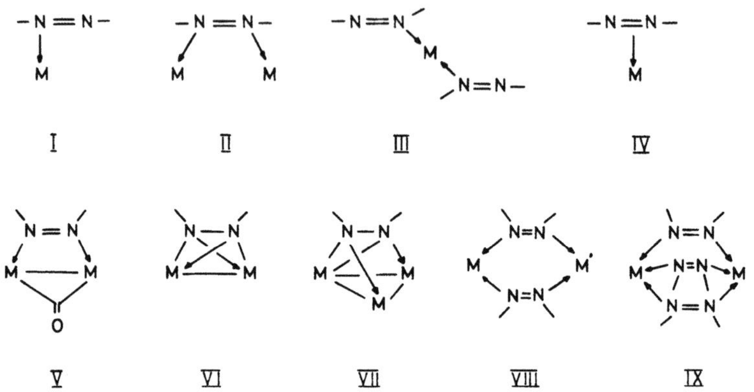

Fig. 3. Possible modes of coordination of the N=N bond to transition metals

The formation of these complexes should be less favoured in cases where the n_1 orbital is strongly delocalized as, *e.g.* in diazirines. It is expected that the bonding interaction between n_1 and the metal decreases the n_1/n_2 interaction, and therefore the n_2 level should be raised in energy. This effect may enable the N=N group to donate the second "lone pair" (n_2) rather than two π-electrons to another metal resulting in formation of the bridged complexes of type II. Depending on the geometry of complex II and the metal involved, metal-metal bond formation may occur to afford complexes V and VI where the diazene group acts as a four-or six-electron donor, respectively. Further possible modes of coordination are the doubly and triply bridged complexes VIII and IX and the cluster compound VII. Finally, π-coordination may be preferred over σ-coordination if the diazene is a poor σ-donor and the metal fragment a poor σ-acceptor.

According to the nature of the HOMO of diazoalkanes one expects the formation of complexes like X or XI by the formal way of a 1.3-dipolar cycloaddition with

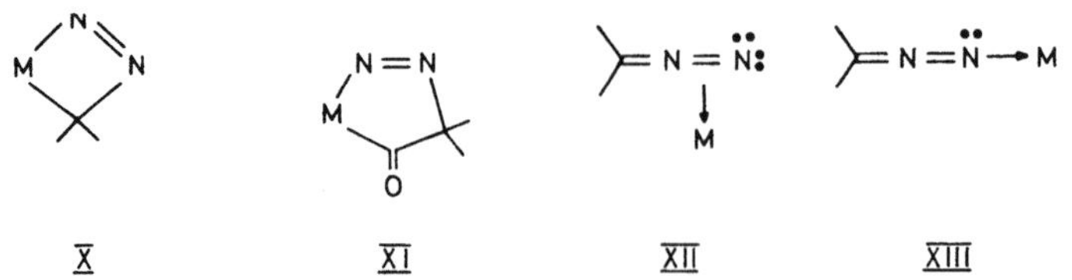

Fig. 4. Possible coordination behaviour of diazo compounds

the metal or an M–L (*e.g.* L=CO) bond. The four-membered metallocycles X should be unstable and eject nitrogen. Side-on coordination of the N=N group (XII) analogous to metal-olefin complexes is probably more likely. σ-donation via a nitrogen lone pair (XIII) can be mentioned, although it would seem unlikely, due to the very low energy of the corresponding orbital[28].

IV. Complexes Derived from Diazenes

1. Mononuclear σ-Complexes

Iron carbonyl complexes are easily prepared by the reaction of $Fe_2(CO)_9$ with a cyclic diazene. An excess of the diazene prevents the more favoured formation of binuclear compounds[29, 30]. Trans-diazenes like trans-dimethyldiazene[31] or azo-

$$Fe_2(CO)_9 + N{=}N \longrightarrow N{=}N{\rightarrow}Fe(CO)_4 + Fe(CO)_5 \qquad (1)$$

Table 3. Physical data of some σ-complexes $LFe(CO)_4$ (type I, Fig. 3)

L		ν(CO)¹)	UV, kK (ε) ¹)	¹H-NMR²)	Ref.
1	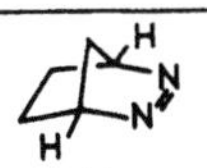	2055, 1977, 1958, 1940	23.0 (1310)	5.65, 5.45 (5.3)	30)
2	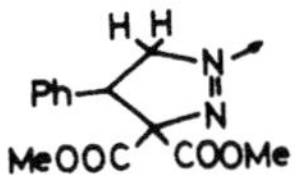	2059, 1983, 1965, 1946	20.8 (1790)	5.4³)	29)
3	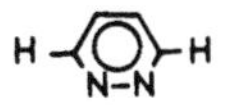	2052, 1972, 1953, 1935	23.3⁴) (1300)	0.6, 1.1⁵) (0.9)	35)
4	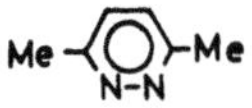	2055, 1976, 1953, 1934	–	7.1, 7.5⁵) (7.47)	35)
5	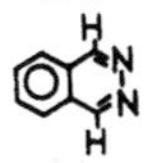	2054, 1982, 1955, 1935	22.3⁶) (1790)	0.8, 0.28⁷) (0.44)	41)
6	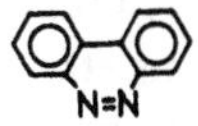	2050, 1970, 1951, 1934	–	–	36)
7	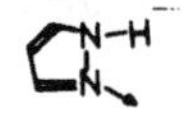	2040, 1950, 1933	–	–	37)
0		2042, 1970, 1943	25.0sh (1250)⁶) 29.0 (1780)	– –	38)

¹) In hexane solution.
²) Toluene-d_8, τ-values of protons indicated in the formula, numbers in parantheses correspond to free ligands.
³) Benzene-d_6.
⁴) In toluene, Ref.[36].
⁵) In THF.
⁶) Ref.[36].
⁷) $CDCl_3$.

benzene[32] do not react according to (1) but yield binuclear complexes VI or bi- and trinuclear complexes containing the o-semidine skeleton, respectively. Use of the corresponding cis-diazenes in this reaction reveals that they are catalytically isomerized to the trans-isomers affording the same binuclear products[33]. The same isomerisation occurs in the presence of $CpMH_2$(M=Mo, W)[34].

From the appearance of four well resolved ν(CO) bands in the IR spectrum of the pyrazoline complex *2* (Table 3) it was initially proposed that the ligand adopts an equatorial position in the trigonal bipyramid[29]. An X-ray structural investigation has revealed that the diazene occupies an apical position, however[39]. The bonding parameters of the pyrazoline are not changed upon coordination in agreement with the formulation of a σ-complex. The coordination geometry around the metal is nearly identical with the pyridine complex *8*[38]; the only difference is the shorter Fe-N bond (1.98 Å) of *2* compared to that of *8* (2.04 Å). It is interesting to note that both compounds have the same rotational orientation, namely, the N=N or N=C bonds are aligned along the Fe-C^1O bond axis (Fig. 5). In contrast to the diazene

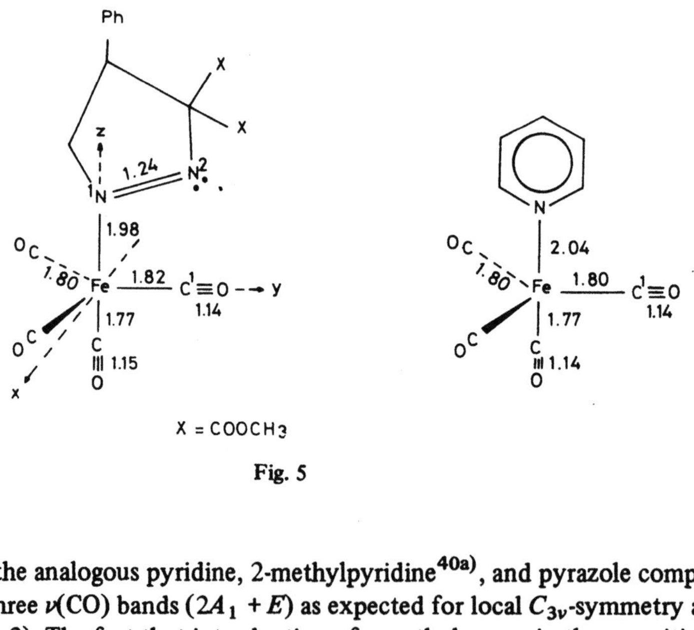

Fig. 5

complexes, the analogous pyridine, 2-methylpyridine[40a], and pyrazole complexes show only three ν(CO) bands ($2A_1 + E$) as expected for local C_{3v}-symmetry at the metal (Table 3). The fact that introduction of a methyl group in the α-position of complexed pyridine does not change the number of CO vibrations, as does introduction of a nitrogen atom, indicates that the appearance of four bands is caused by an electronic effect of the N=N group and not by the unsymmetrical ligand skeleton[40b]. Comparing the wavenumbers of the highest ν(CO) vibration of diazene and aromatic amine complexes (*1–6* and *7, 8*) one recognizes that the latter have a higher electron density at the metal. This points to a somewhat stronger π-back-bonding in the diazene complexes due to the presence of an energetically low π^*_{NN} orbital[42]. Maximum overlap between this orbital and d_{xz} or d_{yz} will occur only if the N=N bond axis is aligned along the *yz* or *xz* plane (Fig. 5). This should give rise to a small energy barrier of rotation around the Fe-N bond and, as a consequence thereof, to the existence of different conformers in solution.

All investigated tetracarbonyliron complexes listed in Table 3 exhibit a low energy electronic absorption band assigned, on the basis of quantum chemical calculations[43], to a predominantly charge-transfer transition from the $Fe(CO)_4$ moiety into the antibonding π^*-level of the N=N bond (CTML). The presence of this absorption band is another distinct difference between (cis-diazene)$Fe(CO)_4$ compounds and amine complexes like *1* and *8*. The dependence of the transition-energy on the nature of the N=N bond is in agreement with this formulation (compare *1* and *2*). This band undergoes a reversible shift to higher energy upon cooling in different solvents to –60 °C; in the case of *2* the deep red colour changes to yellow[30, 36]. This effect and the temperature dependent ^{1}H-NMR spectrum of complex *1* (^{13}C-NMR shows no temperature-effect) may be explained by assuming a conformational equilibrium for this type of complexes due to hindered rotation around the Fe-N bond[36].

Mononuclear σ-complexes of group VIb carbonyls are conveniently prepared by ligand displacement reactions.[44–48] As in the case of iron carbonyls the reaction

(2)

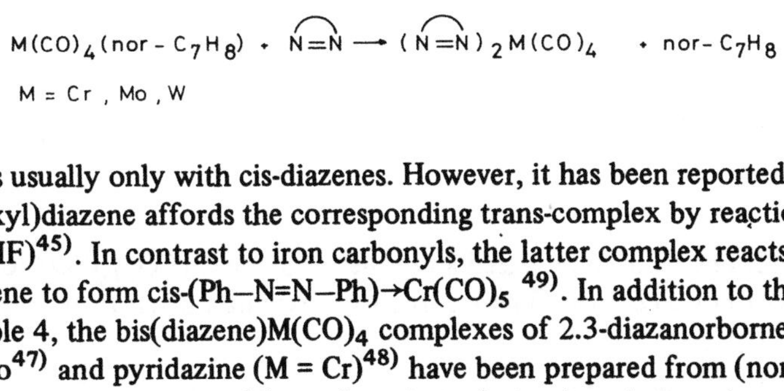

(3)

(2) proceeds usually only with cis-diazenes. However, it has been reported that trans-bis(cyclohexyl)diazene affords the corresponding trans-complex by reaction with $Cr(CO)_5(THF)$[45]. In contrast to iron carbonyls, the latter complex reacts also with cis-azobenzene to form cis-(Ph–N=N–Ph)→$Cr(CO)_5$ [49]. In addition to the complexes listed in Table 4, the bis(diazene)$M(CO)_4$ complexes of 2.3-diazanorbornene (M = Cr[47, 48], Mo[47] and pyridazine (M = Cr)[48] have been prepared from (norbornadiene $Cr(CO)_4$ according to reaction (3); tris(pyridazine)tricarbonyl-chromium was analogously obtained from (C_7H_8) $Cr(CO)_3$ [48]. Beck[46] has reported the preparation of (pentamethylenediazirine)$_2$ $Mo(CO)_3$ from $(C_7H_8)Mo(CO)_3$ and the diazirine; the ligand is assumed to function as a two- and four-electron donor. The same diazirine[46] as well as diazanorbornene[50] react with (methyl-cyclopentadienyl) $Mn(CO)_2(THF)$ or (cyclopentadienyl) $Mn(CO)_2(THF)$ to form the corresponding σ-complexes.

As in the case of (cis-diazene)$Fe(CO)_4$ complexes the diazeno ligand disturbs the local symmetry (C_{4v}) in $L{\cdot}M(CO)_5$ and gives rise to four or five ν(CO) bands instead of three. The low-energy absorption band of (benzo[c]cinnoline) $M(CO)_5$ (M=Cr, Mo, W) was assigned to an $n{\rightarrow}\pi^*$ transition[51]. However, in view of the assignments made for the spectra of $LCr(CO)_5$ compounds (L = amines, phosphines)[52] this transition may be rather of "*d-d*" or CTML character.

The metal-nitrogen bond is easily cleaved by CO, PPh_3, and PF_3 as demonstrated for the benzo[c]cinnoline complexes[51].

Table 4. Physical data of σ-complexes $LM(CO)_5$, (type I, Fig. 3)

L	M	$\nu(CO)_{cm^{-1}}$ [1])	UV, kK (ε)[1])	^{1}H-NMR[2])	Ref.
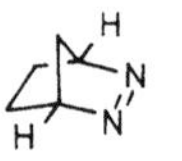	Cr	2073, 1992, 1947, 1929	25.2[3]) (8400)	4.84 4.94	44)
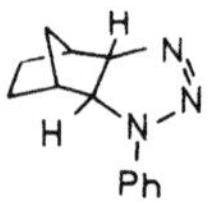	Cr	2071, 1987, 1945, 1939, 1919	(yellow)	5.62 (6.16) 5.46 (5.39)	44)
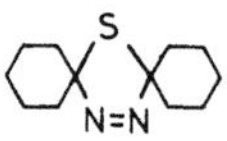	Cr	2073, 1991, 1953, 1941, 1921	(yellow)	–	44)
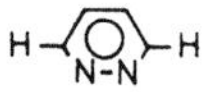	Cr	2074, 1991, 1944, 1919	21.8 (4700)[3]) 24.3 (4300)	0.8 1.2	44)
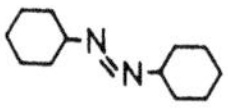	Cr	–	(yellow)	–	45)
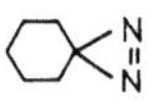	W	2079, 1956, 1943, 1929	(yellow)	–	46)
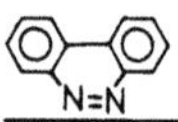	Mo	2074, 1984, 1938, 1896sh	19.0	–	51)

[1]) In n-hexane.
[2]) In acetone-d_6, τ-values of protons indicated in formula, numbers in parentheses correspond to free ligand.
[3]) From Ref.[36]).

Herberhold *et al.*[53]) have prepared a number of arene $Cr(CO)_2$(diazanorbornene) complexes according to Eq. (4). These complexes exhibit a linear relationship be-

$$\mathrm{Ar\,Cr(CO)_3 + N{=}N \xrightarrow[\text{THF}]{h\nu} N{=}N \longrightarrow Cr(CO)_2Ar + CO} \qquad (4)$$

Ar = Me_6C_6, p-$Me_2NC_6H_4Me$, m-$Me_3C_6H_3$, p-$Me_2C_6H_4$, p-$MeOC_6H_4Me$, p-FC_6H_4Me, C_6H_6, p-(MeOOC)C_6H_4Me, p-$(MeOOC)_2C_6H_4$

tween the ν(CO) force constants and the Hammet-constants $\Sigma\sigma_p$. The bridgehead protons of the coordinated diazanorbornene give rise to two signals in the NMR spectrum. Upon heating these signals coalesce and it is assumed that the effect originates from a fast 1.2-shift of the Cr-N bond between the two nitrogen atoms. The temperature of coalescence is lower for electron-rich arene ligands (52–77 °C) than for electron-poor ones (90–95 °C) and $\Delta G^{\ddagger}$ was determined as 17–20 kcal/mole[53]). In this connection it is noted that the N-oxide of diazanorbornene exhibits only one single signal for both bridgehead protons pointing to a symmetrical structure or a fast 1.2-shift of the N–O bond. However, ^{13}C-NMR spectroscopy revealed the presence of nonequivalent bridgehead carbon atoms and therefore firmly established the asymmetry of the N=N–O function[54]. From this it follows that the assignment of symmetrical or unsymmetrical coordination of 2.3-diazanorbornene (and similar diazenes) in the corresponding metal complexes (N=N–M) on the basis of the ^{1}H-NMR spectrum alone might lead to erroneous results.

To gain information as to the donor-acceptor properties of the diazeno ligand, a number of $(Me_6C_6)Cr(CO)_2L$ complexes (L = *e.g.* NH_3, pyridine, pyridazine, 2.3-diazanorbornene, PPh_3; $\nu(CO)$: 1845, 1848, 1855, 1860, 1865 cm^{-1}) were prepared. Comparison of the wavenumbers of the highest $\nu(CO)$ vibration indicates the presence of back-bonding into the antibonding π^*_{NN}-level[53]. These results completely agree with the conclusions obtained on the nature of the Fe-NN bond in the corresponding σ-$Fe(CO)_4$ complexes (Table 3, vide supra).

Vrieze *et al.*[55] have prepared several platinum complexes according to reaction (5). It was shown that the ν(N=N) vibration of the free ligand remains essentially

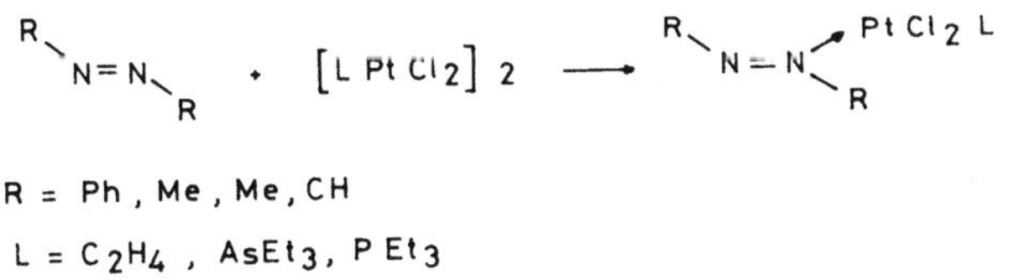

(5)

unchanged upon coordination (IR and Raman spectra); in the case of azobenzene the corresponding values are 1442 and 1448 cm^{-1} for the free and complexed trans-diazene group (L = $AsEt_3$), respectively. The UV spectra of these complexes are very similar to the corresponding piperidine compounds $PtCl_2$(piperidine)L indicating the presence of a "pure" σ-M–N bond.

2. Binuclear σ-Complexes

The first complexes containing two metal carbonyl groups coordinated to the unsubstituted diazeno moiety were prepared by oxidation of complexed hydrazine

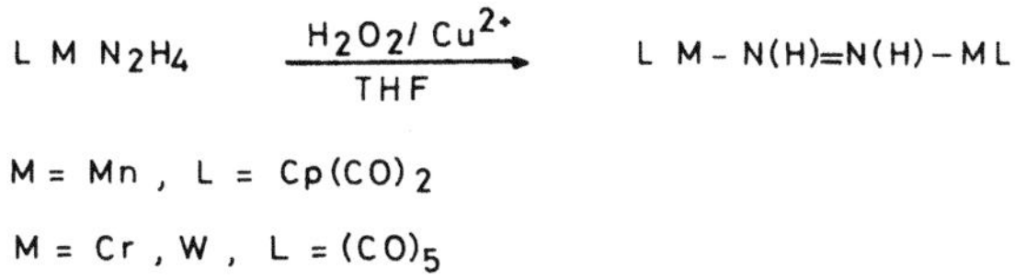

(6)

with H_2O_2/Cu^{2+} or p-$MeC_6H_4SO_2N_3$ by Sellmann *et al.*[56–58]. The IR spectra of the deuterated and ^{15}N-substituted derivatives suggested that the tungsten and chromium complexes in solution probably exist as cis-diazene compounds[57,58]. An X-ray analysis has shown that $[[Cr(CO)_5]_2N_2H_2 \cdot 2\,THF$ contains trans-diazene, however[59]. The nitrogen and chromium atoms lie within the same plane and each

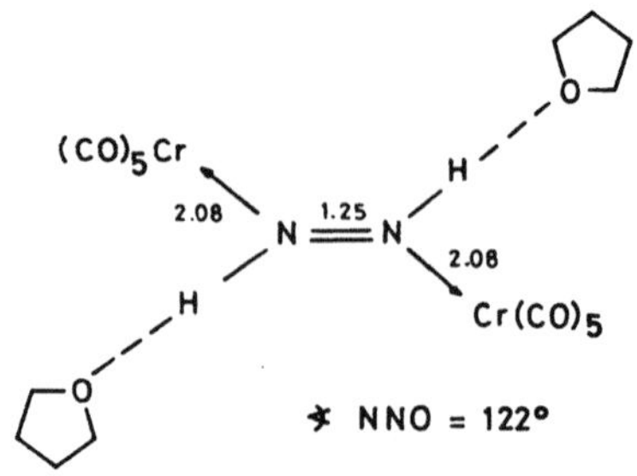

10

Fig. 6

hydrogen of the diazene ligand forms a hydrogen-bond to a THF-molecule which is located 0.5 Å off this plane (N· · · O = 2.98 Å). The relevant bond distances are 1.25 (N=N) and 2.08 Å (Cr–N). From the rather short Cr–N bond a certain amount of π-bonding was inferred[42, 59].

The diimine protons absorb at very low field (–4 to –6τ, Table 5) and are therefore easily exchanged by deuterium ions. In the presence of catalytic amounts of a strong base $N_2H_2[Cr(CO)_5]_2$ irreversibly disproportionates to $N_2[Cr(CO)_5]_2$ and $N_2H_4[Cr(CO)_5]_2$ [60].

Bridged binuclear σ-complexes containing the cis-diazene group fixed in a three-membered ring were prepared from diazirines by reaction with $Fe_2(CO)_9$ [19] or

Table 5. Physical data of binuclear σ-complexes $L \cdot [M(CO)_nL']_2$ (type II, Fig. 3)

	L	$M(CO)_nL'$	ν(CO), cm^{-1} [1]	Colour[2]	^{1}H-NMR[3]	Ref.
9	HN=NH	$CpMn(CO)_2$	1880, 1915[4]	blue	–4.27 (NH)	[56]
10	HN=NH	$Cr(CO)_5$	2055, 1950[5] 1923	violet	–6.14 (NH)	[57]
11	HN=NH	$W(CO)_5$	–	violet	–6.38 (NH)	[58]
12		$Fe(CO)_4$	2083, 2045, 2014 2005, 1985, 1961	violet	9.5 (Me)[6]	[19]
13		$Cr(CO)_5$	–	red	9.65[6]	[36]
14	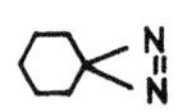	$W(CO)_5$	2059, 1972, 1945, 1930	violet	–	[46]
15	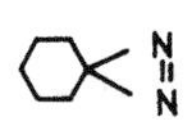	$Fe(CO)_4$	2082, 2045, 2015 1998, 1985, 1973	violet	9.3, 8.8[6] (CH_2)	[19, 46]

[1]) In n-hexane.
[2]) in hydrocarbon solution.
[3]) τ-Values, acetone-d_6.
[4]) In benzene.
[5]) In THF.
[6]) In benzene-d_6.

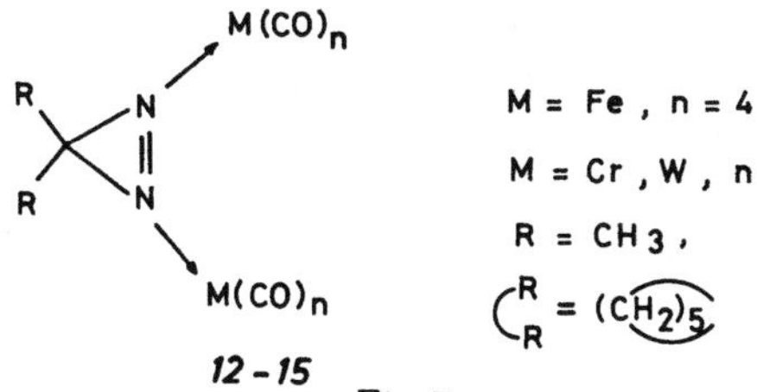

12–15

Fig. 7

$M(CO)_5(THF)$, M=Cr, W,[36, 46] (*12–15*). In contrast to the diimine complexes, compounds *12–15* show a more complicated $\nu(CO)$ pattern due to steric interference of the two cis-$M(CO)_n$ groups. The deep-red to violet colour of these complexes in solution seems to be a general property of the $M \leftarrow N(R) = N(R) \rightarrow M$ moiety.[61]

Table 6. Physical data of complexes $LFe_2(CO)_7$ (type V, Fig. 3)[1)]

L		$\nu(CO)$, cm^{-1} [2)]	Colour[2)]	^1H-NMR[3)]
16	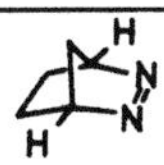	2057, 2004, 1984 1969, 1802	red	5.32 (4.98)[4)]
17	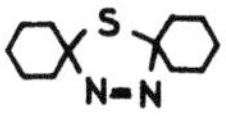	2062, 2009, 1996 1963	green	–
18	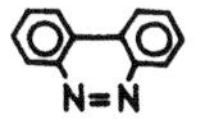	2049, 2004, 1982 1963, 1972	yellow-green	–
19	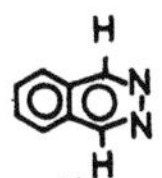	2047, 1995, 1973[5)] 1950, 1936sh, 1775	red	0.62 (0.57)
20	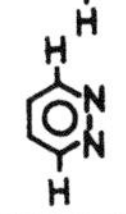	2040, 1989, 1968 1949, 1775	blue-green	1.21 (1.0)

[1)] From Ref.[62)].
[2)] In THF solution.
[3)] In acetone-d_6, τ-values of protons indicated in formula.
[4)] Values in parentheses correspond to free ligand.
[5)] See also Ref.[41)].

Herberhold[62)] and Alper[41)] obtained another type of bridged complex (V, Fig. 3) from the reaction of $Fe_2(CO)_9$ with cyclic diazenes (Table 6). On the basis of IR and ^1H-NMR spectra a symmetrical $L \cdot Fe_2(CO)_7$ structure containing an Fe-Fe bond and a bridging diazene and CO group was assigned to these complexes. Except the complexes *19, 20*, these compounds are easily decarbonylated to yield $L \cdot Fe_2(CO)_6$ complexes in which, in addition to the two lone pairs, the diazeno ligand also donates its π-electrons to the metals. In the case of pyridazine and phthalazine ligands (*19, 20*) the π(N=N)-orbital is part of the aromatic ring system which would be destroyed if this orbital is used in bonding to a metal[62)].

In addition to mono bridged complexes a number of doubly bridged compounds (type VIII, Fig. 3) were prepared by reaction of diazenes with (nor-C_7H_8)$M(CO)_4$, M=Cr, Mo, W[36, 46–48]. Similar complexes of type IX containing three bridging diazeno ligands were obtained from the reaction of $Cr(CO)_6$ with diazanorbornene or pyridazine in boiling dioxane[48].

Table 7. Physical data of doubly and triply bridged binuclear complexes (types VIII and IX, Fig. 3)

	Complex		ν(CO), cm^{-1} [1]	Colour	^{1}H-NMR[2]	Ref.
21	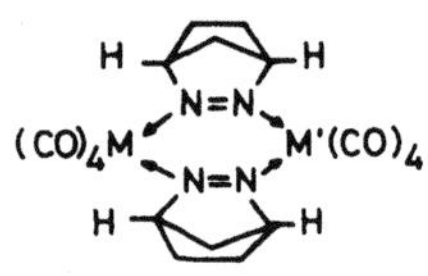	M = M' = Mo	2043, 2010, 1948 1920, 1882	red-brown	4.9[3]	47)
		M = M' = Cr	2033, 1997, 1939[4] 1913, 1898sh, 1877	brown	4.7	48)
		M = Mo, M' = Cr	2035, 2005, 1940 1912, 1872	black	4.87[3]	47)
22	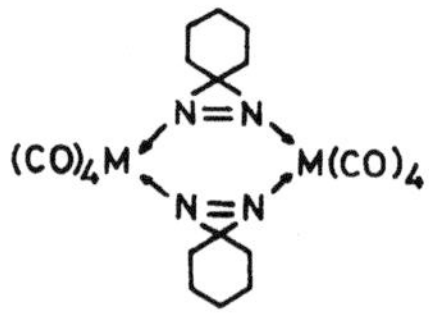	M = Mo M = Cr	2014, 1967, 1948[5] 1942, 1910	–	–	46) 36)
		M = W	2006, 1959, 1930[5] 1890	–	–	46)
23	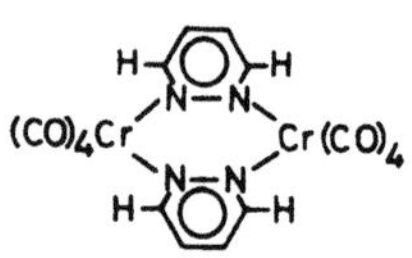		1976, 1926, 1874[6] 1828	black	0.86	48)
24	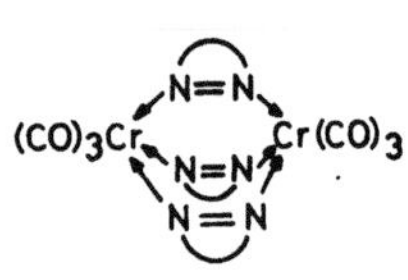	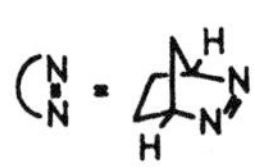	1897, 1844[4]	dark-green	5.11	48)
		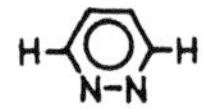	1893, 1811[6]	green-black	0.33	48)

[1]) In chloroform.
[2]) In acetone-d_6, τ-values of protons, indicated at the left.
[3]) In $CDCl_3$.
[4]) In benzene.
[5]) In n-hexane.
[6]) In THF.

3. π-Complexes

In 1971, Otsuka *et al.* [63, 64], and Klein and Nixon[65] simultaneously reported the formation of nickel π-complexes of azobenzene. The π-bonded structure (type IV, Fig. 3) was assigned on the basis of displacement reactions and spectroscopic evidence. Treatment of (π-Ph–N=N–Ph)Ni $(PR_3)_2$, (R = Me, Bu, Ph), with aqueous ethanol or with dimethylglyoxime in THF resulted in the reduction of the diazeno ligand to hydrazobenzene[65]. Analogous π-complexes of Co, Mo, Ti and V were recently reported[66–69] (see Table 8).

The proposed structure was confirmed by X-ray analysis of $L_2Ni(PhN{=}NPh)$, L = Me_3CNC, $P(p\text{-}MeC_6H_4)_3$, by Ibers *et al.*[70–72]. If the diazene is regarded as a single ligand, the Ni atom has its expected trigonal coordination. Comparison with the

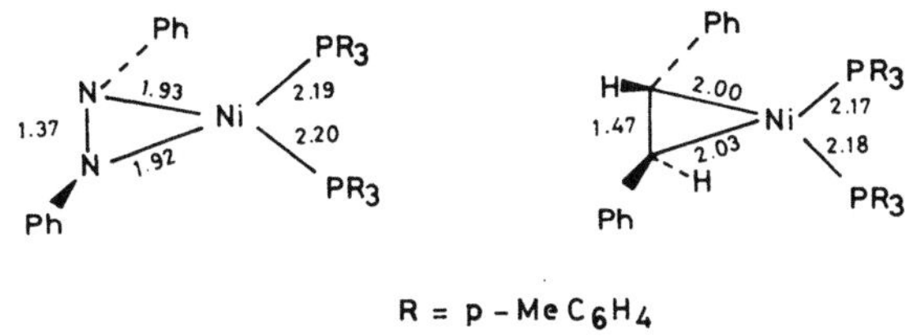

Fig. 8

isoelectronic complex of trans-stilbene[73] reveals that the phenyl groups are bent back further in the diazene complex, indicating better back-bonding. This is further strongly supported by the long N–N bond distance of 1.37 Å compared to 1.24 Å in free azobenzene[73, 74]. Accordingly, the ν(N=N) vibration[75] is lowered by almost 600 cm^{-1} upon coordination while the corresponding shift of ν(C=C) of ethylene on coordination to nickel[70] is 378 cm^{-1}. The π^*-orbital in azobenzene should be at much lower energy than in trans-stilbene and one expects that more electron density is transferred to azobenzene via back-bonding. This effect should become stronger with increasing basicity of the coordinated phosphines accumulating negative charge on the diazeno ligand. Thus, introduction of more basic phosphines shifts the low-energy electronic adsorption band ($n \to \pi^*$?) in the direction of the spectrum of the dianion of azobenzene[71] (Table 8).

Diazenes with carbonyl groups like bis(benzoyl)diazene prefer coordination at oxygen and nitrogen[76] affording metallocycles *25*. However, $(CO)_2Mo(S_2CNEt_2)_2$ probably also forms a π-complex.[68]

4. σ/π-Complexes, Cluster Compounds

The most stable, and probably therefore, first detected metal carbonyl complexes of diazenes are the binuclear compounds (cis-diazene) $Fe_2(CO)_6$, reported by Bennett[77] in 1970. They are conveniently prepared by irradiating or heating $Fe(CO)_5$ in the presence of a cyclic diazene. The structure of several complexes has been confirmed by X-ray analysis and was reviewed recently[2b, c]. In addition to the four lone-pair electrons, the diazene donates two π-electrons to the metal. Accordingly, the N=N bond is lengthened from 1.24 to 1.38 Å. If both nitrogen atoms are regarded as 3-electron donors these complexes represent a further example of the type $Fe_2(CO)_6$ (X) (Y) where X and Y are groups like Br, I, NR_2, PR_2, AsR_2, SR. Table 9 contains physical properties of some selected complexes.

Table 8. Some properties of π-complexes (RN=NR) ML_2, (type IV, Fig. 3)

M,	R,	L	UV, kk[1])	Preparation	Ref.
Ni	Ph	PEt_3	20.55 (23.8)	$Ni(COD)_2$+PEt_3+PhN=NPh	70)
Ni	p-FC_6H_4	PEt_3	20.5 (24.1)		
Ni	p-MeC_6H_4	PEt_3	20.3 (23.3)		
Ni	p-$EtOC_6H_4$	PEt_3	20.3 (23.2)		
Ni	p-$H_2NC_6H_4$	PEt_3	20.8 (23.2)		
Ni	Ph	PPh_3	19.0	$(PPh_3)_2NiCl_2$ + PhN=NPh $\xrightarrow[THF]{Li}$	71, 65)
Ni	Ph	$P(p\text{-}MeC_6H_4)_3$	19.2 (23.0)	"$[P(p\text{-}MeC_6H_4)_3]_2Ni$" + PhN = NPh	71, 64)
Ni	Ph	PPh_2Et	19.6		71)
Ni	Ph	$PPhEt_2$	20.0		
Ni	Ph	PMe_3	20.45		
Ni	Ph	PBu_3	20.75		
Ni	Ph	Me_3C–NC	21.5		
Ph–N=N–Ph			22.7		
Li^+(Ph–N=N–Ph)$^-$			16.5		
$(Li^+)_2$(Ph–N=N–Ph)$^{2-}$			26.3		
Co	Ph	$P(Me)_3$	–	$(PMe_3)_4$ Co + Ph–N=N–Ph	66)
Mo	Ph	Cp	–	Cp_2MoH_2 + PhN=NPh	69)
V	Ph	Cp	–	$Cp_2M(CO)_2$ + Ph–N=N–Ph, M = Ti, V	67)
Mo	COOEt	S_2CNEt_2	–	$(CO)_2Mo(S_2CNEt_2)_2$ + EtOOCN = NCOOEt	68)

[1]) In toluene, numbers in parentheses refer to $n \rightarrow \pi^*$ transitions of free ligands.

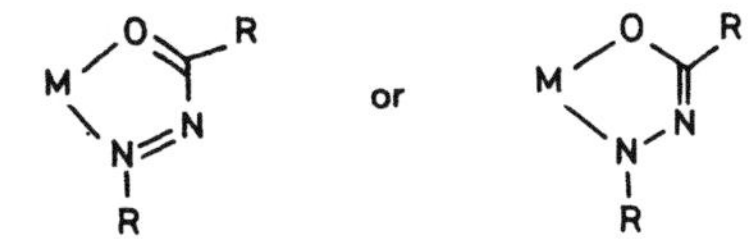

25

R = e.g. : OMe , Ph , C(O)Ph

Fig. 9

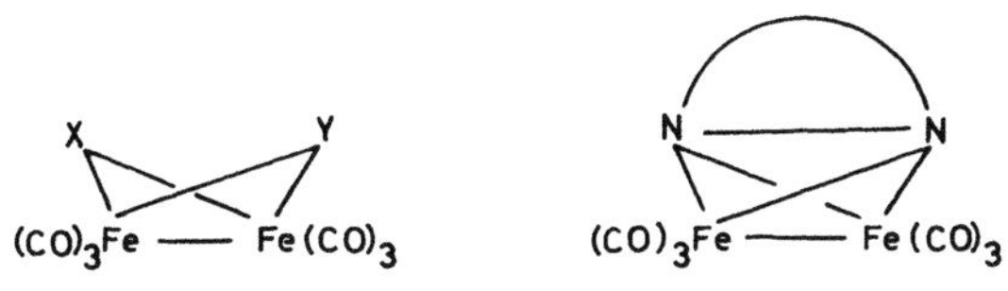

VI

Fig. 10

Table 9. Physical data of binuclear complexes L · $Fe_2(CO)_6$ (type VI, Fig. 3)

		ν(CO), cm^{-1} [1]	UV, kK (ε · 10^{-3}) [2]	^{1}H-NMR[3]	Ref.
27	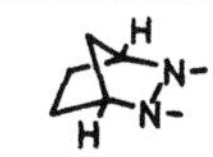	2068, 2022, 1991 1970, 1959	25.0 sh (1.8), 29.0 sh (2.4) 33.6 (30)	7.2 (5.34)	50, 77)
28	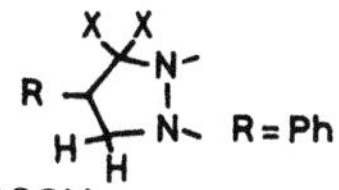 R = Ph X = COOMe	2074, 2033, 1993 1985, 1971	24.0 sh (1.7), 29.0 sh (3.5) 34.0 (24)	7.3–6.3 (5.4)	78)
	R = Me_2CH	2073, 2031, 1992 1983, 1971	24.0 sh (1.7), 29.0 sh (3.0) 33.7 (25.7)	6.9–6.7 (5.4–6)	78)
	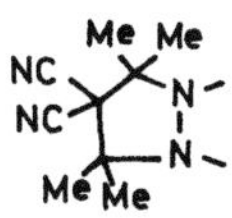	2075, 2031, 2000 1988, 1974	24.0 sh (1.75), 28.5 sh (2.8) 33.5 (33)	9.03 (8.24)	36)
29	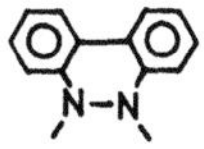	2069, 2030, 1989 1985, 1972	22.0 sh (1.3), 29.0 sh (11) 31 (26.3), 34.5 (22.7)	–	77)
	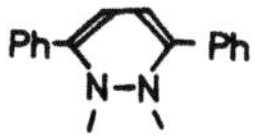	2065, 2025, 1985 1970	–	–	79)
	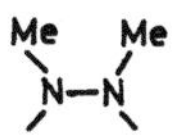	2070, 2024, 1989 1972, 1962	23.5 sh (1.0), 29.2 sh (3.2)[4] 35.1	6.85[5]	31)

[1]) In n-hexane.
[2]) In n-hexane, Ref.[80]).
[3]) In benzene-d_6, τ-values of protons indicated at the left, numbers in parentheses correspond to free ligands.
[4]) Ref.[36]).
[5]) In $CDCl_3$.

Reinvestigation[30] of the reaction of diazanorbornene with $Fe(CO)_5$ led to the isolation of an interesting (cis-diazene) $Fe_3(CO)_9$ cluster of proposed structure *26*. The same type of clusters may be prepared from $Ru_3(CO)_{12}$ and cyclic diazenes[81].

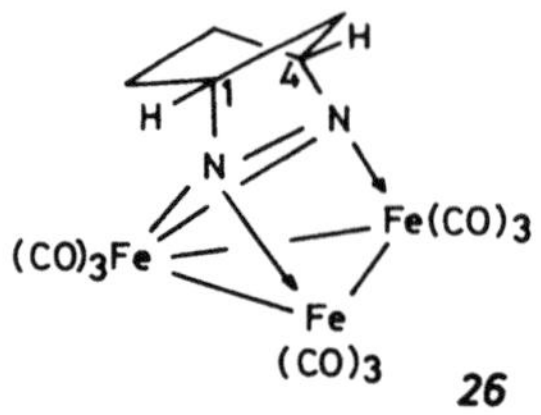

26

^1H, ^4H : 7.0τ (C_7D_8)
^1C, ^4C : 66.6 ppm ($CDCl_3$)

Fig. 11

5. Complexes Formed by Cleavage of the N=N Bond.

When investigating the modes of coordination of cyclic diazenes to iron carbonyls, it seemed interesting to use diazirines, the cyclic isomers of diazoalkanes as ligands.[19] From the known chemistry of these compounds it was expected that, in addition to complexes containing the intact diazirine ring, formation of dinitrogen, carbene or diazoalkane complexes might occur via cleavage of the C–N bonds[82, 83]. Surprisingly, not the C–N bond but the N=N bond is cleaved during reactions of diazirines with $Fe_2(CO)_9$ or $Fe(CO)_5/h\nu$ in hydrocarbon solvents. If the reaction is conducted with $Fe_2(CO)_9$ in THF, the bis-(N,N')-tetracarbonyliron complex *12* is obtained as the major product as well as minor amounts of the cluster *30* and the ketiminato complexes *31* and *32*.

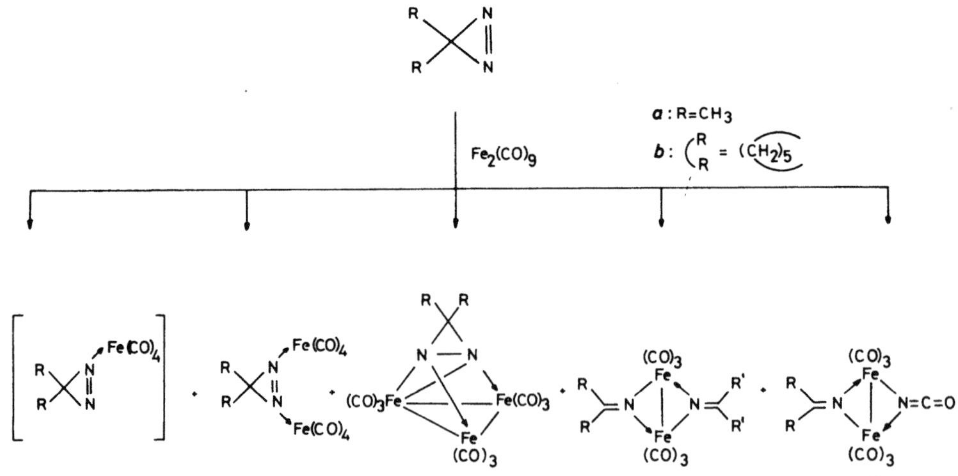

12 *30* *31* *32*

Fig. 12

IR spectroscopic analysis of the reaction of 3.3-dimethyldiazirine with ennea carbonyldiiron in n-hexane indicated the initial formation of a labile tetracarbonyl complex which is converted via *12* into predominantly *31* and *32*. Solutions of *12* decompose into *31* and *32*; in the presence of a different diazirine, R'_2CN_2, the mixed complexes of *31* ($R \neq R'$) are obtained as additional products. The formation of the ketiminato complexes *31* requires the reaction of two diazirine molecules and loss of two nitrogen atoms. These are not lost as gaseous nitrogen but as an isocyanate containing very insoluble material of unknown structure. Thus, *12* seems to be a key intermediate in these reactions. Opening of one C–N bond, insertion of nitrogen into a Fe–CO bond, cleavage of the N–N bond and further rearrangement gives the stable, doubly bridged isocyanate complex *32*; in the presence of excess diazirine the bis(ketiminato) complex *31* is formed, probably by a similar pathway.

The structures of compounds *31* and *32* have been confirmed by X-ray analysis[85], those of *12* (Table 5) and *30* are based on spectroscopic data (Table 10).

Table 10. Physical data of complexes derived from diazirines and iron carbonyls

Compound	ν(CO)(cm^{-1}) [1]	^{1}H-NMR (τ)[2]	UV kK (log ϵ)[1]
12b	2082 m, 2045, 2015 s, 1998 1985, 1973 all vs	9.3 m, 8.8 m	
30a	2074 s, 2020 (sh), 2018 vs 1989 s, 1972 m, 1967 m	9.03 s, 8.59 s	18.1 (3.2) 23.9 (sh) (3.58)
31a	2062, 2020, 1985, 1967 all vs 1960 s	8.25 s	25.6 (sh) (3.52) 30.8 (3.82)
31b	2060, 2020, 1985, 1970 all vs 1960 s	7.75 m, 8.75 m	24.6 (3.54), 30.8 (3.82)
32a	2208 s (νNCO); 2076, 2043 2007, 1992 all vs	8.44 s	20.2 (3.28), 28 (3.85) 32.5 (3.86)
32b	2210 s, (νNCO); 2076, 2043 2007, 1993 all vs	7.9 m, 8.85 m	20.1 (3.23), 27.8 (3.88) 32.7 (3.84)

[1]) In n-hexane.
[2]) In benzene-d_6; all values from Ref.[19]).

Simultaneously with this work, Beck and coworkers have prepared also the complexes *12b*, *31b* and *32b*[46]).

The cleavage of the N=N double bond is new in the chemistry of diazirines. It may be compared with the azobenzene-o-semidine rearrangement; occurring in the reaction of iron carbonyls with azobenzene[2a]), the reaction of bis(trifluoromethyl) diazene and trans-ClIr(CO)($PMePh_2$)$_2$ to yield the nitrene complexes (Ph_2PMe)$_2$Ir(CO)(Cl)(NCF_3)[86]), and with the formation of the ketimine complex CpMn(CO)$_2$(HN = CMePh) from CpMn(CO)$_2$(THF) and 1-diazo-1-phenyl-ethane[87, 88]). In contrast to the behaviour of iron carbonyls, no cleavage of the N=N bond was observed in reactions with group VIb carbonyls[46, 19]).

6. Summary

The results obtained from cyclic diazenes and iron carbonyls serve as an example to discuss the modes of coordination of an isolated N=N bond[30, 62, 46, 19]. As predicted from the electronic structure of these ligands, σ-coordination via the n_1 orbital is usually favoured over π-coordination. Only if the metal is a very good π-donor (see Table 8) and the diazene a good π-acceptor, like azobenzene, will stable π-complexes be favoured. Thus the first reaction step usually affords the σ-N-tetracarbonyl

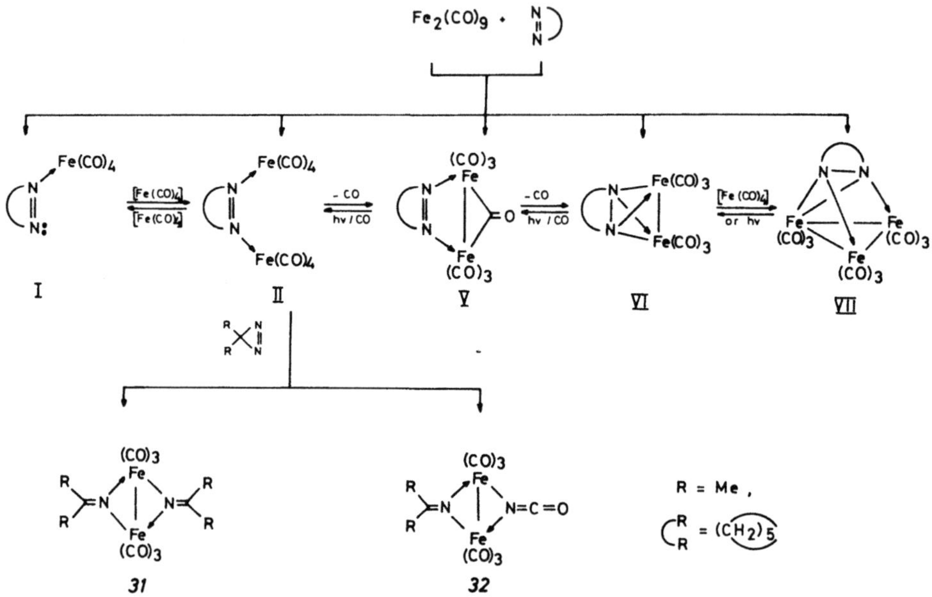

Fig. 13

complexes of type I. These are very unstable in the case of diazirines, probably due to the relatively high ionisation potential (Table 1) and the very low lone pair character of the n_1 (HOMO) orbital (Table 2). The second step gives the N,N'-bis(tetracarbonyliron) complexes II, which were isolated only from diazirines and are assumed to be intermediates in reactions of five- and six-membered diazenes. In this latter case they are easily decarbonylated to the complexes V and VI. This reaction path is very unfavourable in the case of diazirines as demonstrated by the absence of complexes of type V and VI among the reaction products. The different stability and reactivity of compounds of type II is obviously a function of the ring size of the cyclic diazene involved. The distance of the centers of charge of both nitrogen lone pairs should increase considerably in going from six- to three-membered rings; assuming a planar Fe ← N=N → Fe arrangement for these complexes, the corresponding Fe–Fe distances should be 3.2, 3.5, 4.0 and 4.7 Å for a six, five, four and three-membered diazeno ligand, respectively. Thus, diazirines offer optimal steric conditions for the accomodation of one $Fe(CO)_4$ group at each nitrogen but pose difficulties for the decarbonylation reactions producing V and VI, which would require bond formation between two widely separated metal carbonyl groups. In this case therefore complexes of type II prefer a new reaction path involving cleavage of N=N and C–N bonds to yield the more stable doubly-bridged complexes *31* and *32*. The driving force of all these reactions seems to be the formation of the most stable $Fe_2(CO)_6$ complexes.

From the fact that trans-diazenes do not form any of the complexes described, it may be concluded that the initial interaction of the N=N bond with the metal is largely determined by the symmetry of the HOMO (n_1)[89]. Both trans and cis-diazenes possess similar ionisation potentials but only the latter have a HOMO of proper symmetry to overlap with the LUMO (d_{xy}, d_{xz}, d_{yz} or $d_{x^2-y^2}$) of the reacting iron carbonyl species.

Group VIb metal carbonyls[44–46, 51] afford some additional types of complexes which are summarized in Fig. 14.

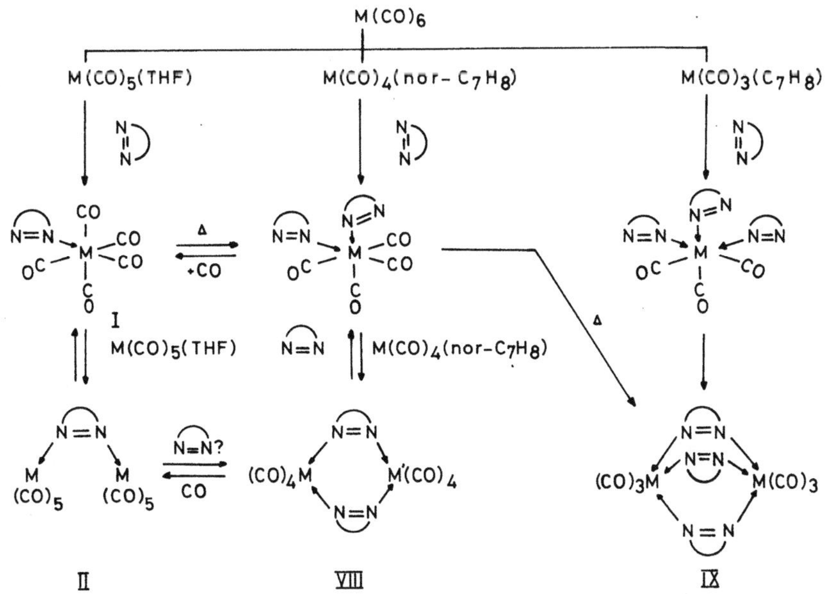

Fig. 14

Here again the scheme applies only for cis-diazenes. However, as already mentioned, trans-bis(cyclohexyl)diazene yields the corresponding (trans-diazene) $Cr(CO)_5$ complex by reaction with $Cr(CO)_5THF$[45]. Another interesting complex containing two different metals bound to one N=N group (VIII, $M \neq M'$, M = Cr, M' = Mo) was prepared from bis(diazanorbornene)$Cr(CO)_4$ and (nor-C_7H_8)$Mo(CO)_4$ by Ackermann and Kou[47].

Aromatic trans-diazenes easily form orthometallated complexes, sometimes accompanied by cleavage of the N=N bond to give the o-semidine containing products. These reactions have been recently reviewed by Bruce and Goodall[2a].

V. Reactions of Diazenes Activated by Iron Carbonyls

1. Reactions of Alkynes with (N–N)$Fe_2(CO)_6$

One of the major topics of organometallic chemistry is the activation of organic functions by coordination to a metal. The reactivity of the complexed diazene group was recognized when heating the (pyrazoline)$Fe_2(CO)_6$ complex *28a* in the presence of excess diphenylacetylene.[80, 91] Surprisingly, (tetraphenylcyclobutadiene)tricarbonyliron (*33*) is formed as the major product (70% yield) among decomposition products of the pyrazoline, small amounts of hexaphenylbenzene and known complexes of the alkyne. If the reaction is conducted photochemically at room temperature, three new complexes, *34c*, *35c* and *36i* were obtained. The structure of *34c* was resolved by X-ray analysis[91]. The formation of *34c* may be viewed as an insertion of the alkyne into one Fe–N bond and π-coordination of the double bond formed to one iron atom. Cycloadduct *34c* forms black crystals which decompose in solution into

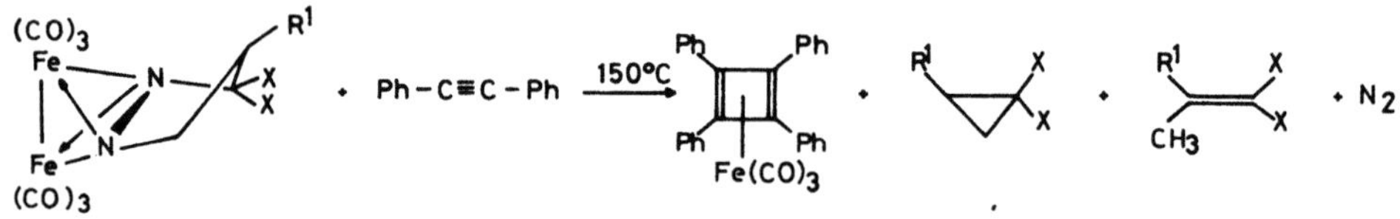

28a *33*

R^1 = Ph, X = $COOCH_3$

Fig. 15

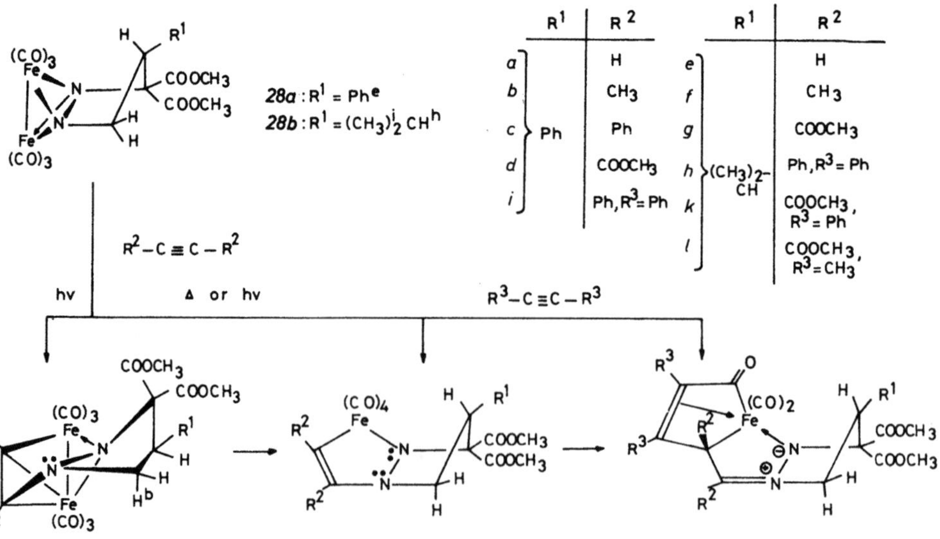

34c *35a,c,d,e,f,g* *36i,h,k,l*

Fig. 16

pink coloured complex *35c*. The "double-addition" product *36i* contains a second molecule of the alkyne and is the most stable complex in this series. However, heating *36i* in a vacuum to 150 °C affords (tetraphenylcyclobutadiene)tricarbonyliron. This observation and the fact that *35c* is converted into *36i* by reaction with further diphenylacetylene suggest that these compounds are intermediates in the cyclodimerisation of the alkyne. This was confirmed by IR analysis of the reaction of *28a* and diphenylacetylene at 150 °C. The proposed structure of *36i* was verified by X-ray analysis[85] (Fig. 20). Using other alkynes R C_2R (R = H, CH_3, $COOCH_3$) in these reactions led to isolation of a number of further complexes of types *35* and *36*.

To investigate the influence of different diazeno ligands on these reactions, the diazanorbornene and benzo [c] cinnoline complexes *27* and *29* were also reacted with a variety of alkynes. The reactivity of *27* (Fig. 17) is similar to that of *28*. The

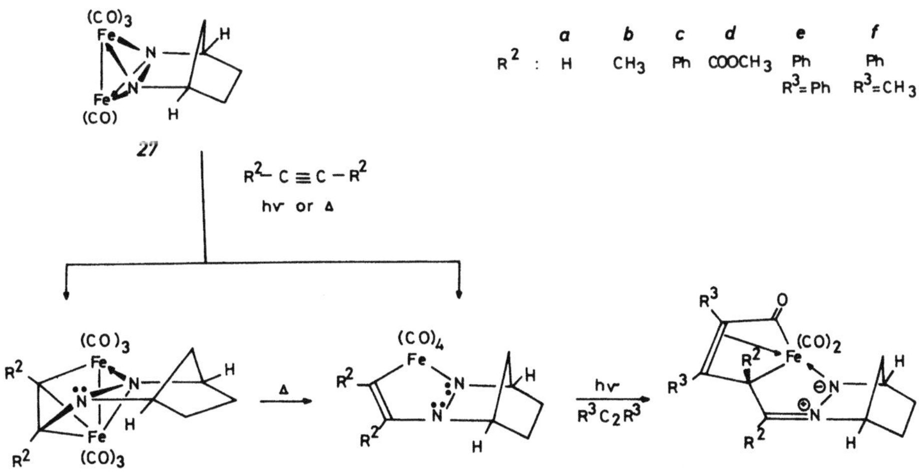

Fig. 17

structure proposed for metallocycles *35* and *38* is based on the spectroscopic data of *38a* (R^2 = H). The bridgehead protons absorb at 6.17 and 5.33 τ (benzene-d_6) and the corresponding carbon atoms at 73.6 and 64.9 ppm (toluene-d_6)[92]. These values are similar to those of free diazanorbornene (5.3 τ and 75.3 ppm) implying some double-bond character for the N–N bond. The very low chemical shift of the vinylic hydrogens (0.75 and 2.87 τ) points to a certain amount of electron delocalisation within the five-membered metallocycle (Fig. 18); these compounds are therefore formulated as diazaferroles, analogous to the previously reported diazamolybdoles[93].

In contrast to the pyrazoline and diazanorbornene-$Fe_2(CO)_6$ complexes (*27, 28*), the benzo[c]cinnoline compound *29* yields only the cycloadducts *40* and no diazaferrole or "double-addition" products (Fig. 19).

Attempts to convert *40* into the latter complexes by heating have been unsuccessful. This characteristic difference in chemical reactivity may be rationalized by

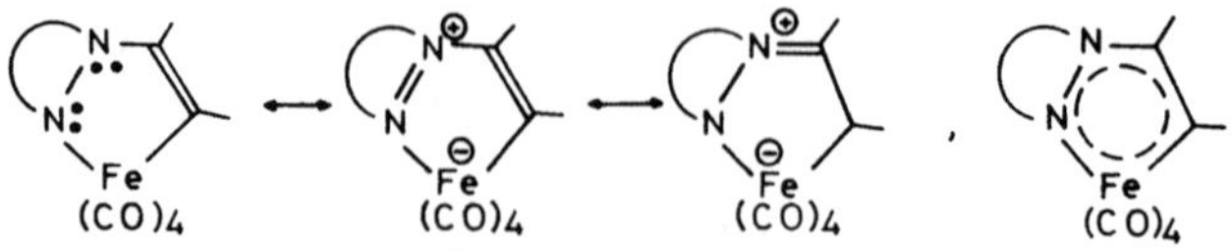

Fig. 18

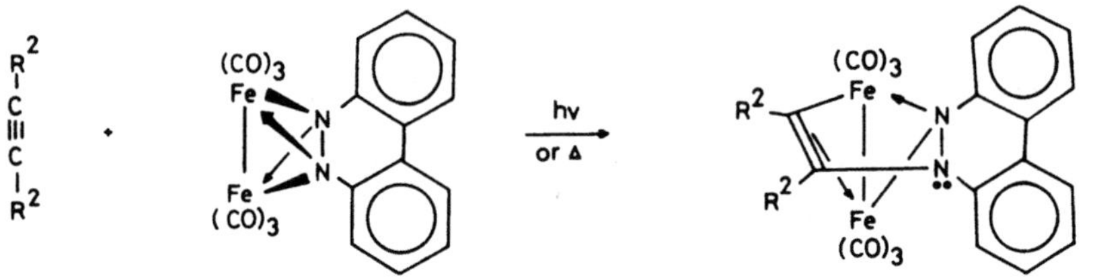

39 *40 b,c,d*

R^2 : CH_3, Ph, $COOCH_3$

Fig. 19

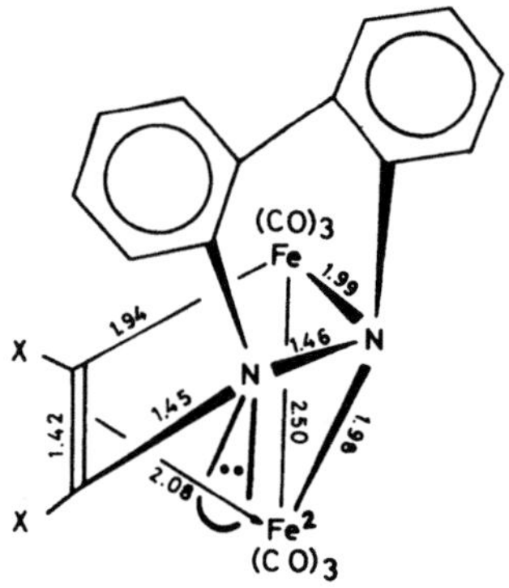

40 d

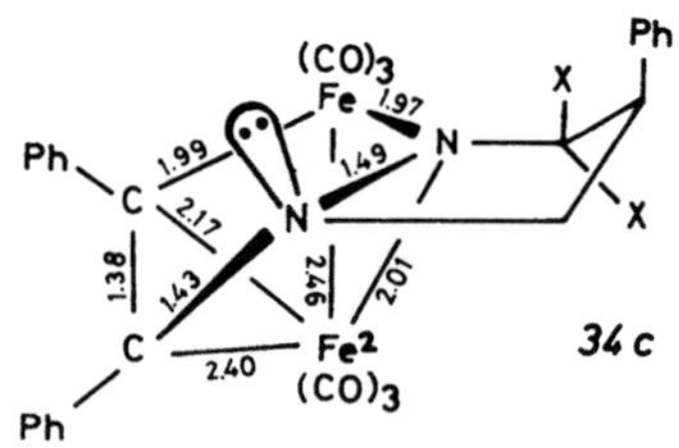

34 c

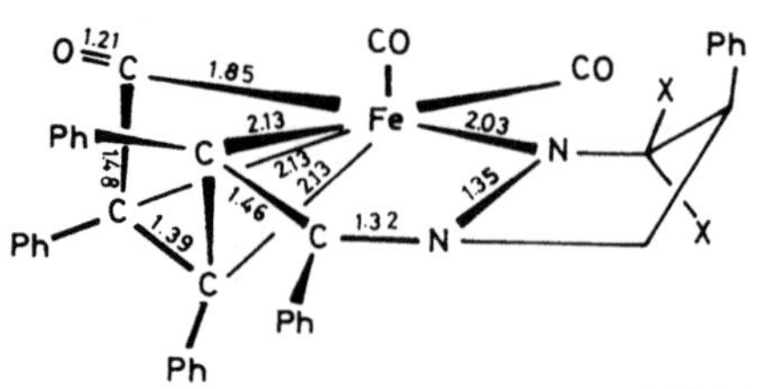

36 i

X = $COOCH_3$

Fig. 20

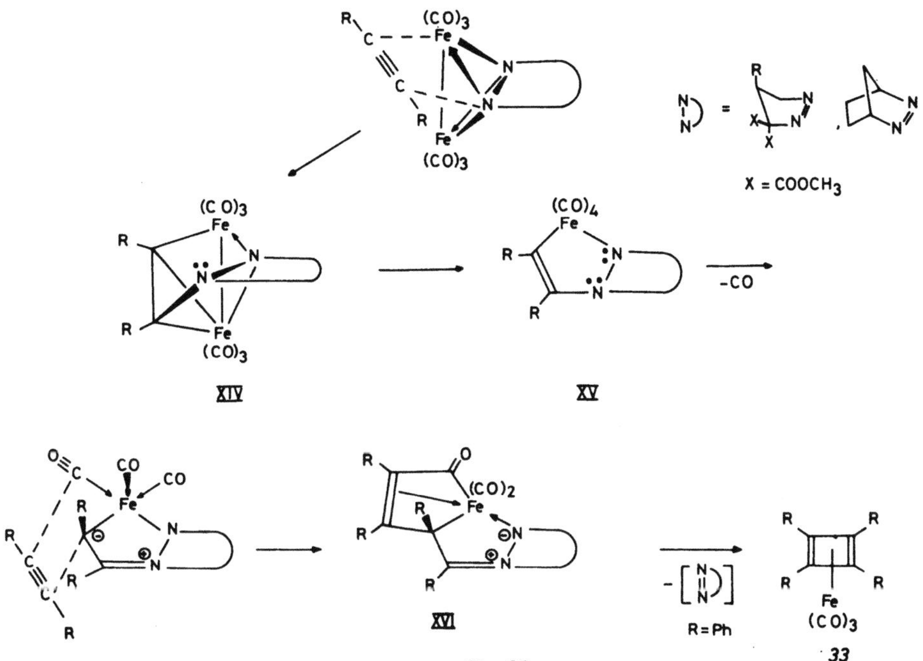

Fig. 21

the assumption that loss of the $Fe^2(CO)_3$ moiety (Fig. 20), which is required for the formation of the diazaferrole, is initiated by nucleophilic attack of CO, solvent, or another molecule of *40* on Fe^2. The major structural difference between *34c* and *40d* is the orientation of the lone pair at N^1 with respect to Fe^2. In *34c* this orbital is directed opposite to Fe^2, whereas in *40d* it effectively shields the $Fe^2(CO)_3$ group and thus prevents nucleophilic attack at the metal. The main features of the molecular structures of *34c, 40d* and *36i* are shown in Fig. 20[85, 91].

Fig. 21 summarizes the complexes obtained. The formation of cycloadducts XIV can be explained formally as insertion into the Fe–N bond. By analogy to the well known cyclo-oligomerisation reactions of acetylenes[94] in the presence of transition metals, it seems probable that the reaction occurs stepwise and is initiated by formation of a π-complex of the alkyne. The necessary vacant coordination site may be formed by loss of CO or by opening of one Fe–N bond.

Complexes of type XIV are more stable in the case of diazanorbornene derivatives (*37*) than in the pyrazoline compounds where only one cycloadduct (*34c*) could be isolated. The characteristic property of XIV is the smooth transformation into the diazaferroles XV which can easily be monitored by ^1H-NMR spectroscopy. This suggests that the leaving iron carbonyl fragment is converted into a diamagnetic complex.

Inhibition experiments with carbon monoxide reveal that the introduction of the second alkyne is initiated by loss of CO. The alkyne occupies the vacant coordination site and finally yields the double-addition product XVI by forming two C–C bonds. This final step resembles the 1.3-dipolar cycloaddition of alkynes and 1.3-dipoles[95]. This mechanism is discussed in detail in Section V. 2.

Table 11. IR and electronic spectra of complexes *34–40* [1])

34c	2067m, 2053m, 2020sh, 2013s, 1966m [2])	1730m	16.5 (0.7), 22.5sh (2.9), 29.2 (7.7) [3])
37a	2054s, 2010vs, 1975vs, 1954vs, 1942w		17.0 (0.74), 24.5sh (2.56), 29.6 (9.2) [3])
37b	2047, 2001, 1968, 1947all vs, 1936s		17.0 (0.58), 24.0sh (2.2), 29.3 (9.3) [3])
37c	2050s, 2008vs, 1976vs, 1956w, 1948vs, 1937w		16.3 (0.6), 24.7sh (2.5), 29.4 (7.9) [3])
40b	2065s, 2032vs, 1990vs, 1976m, 1964w		22.3 (3.9), 31.1sh (9) [3])
40c	2066s, 2040vs, 1999vs, 1992vs, 1985m, 1952w		22.1 (3.6) [3])
40d	2082s, 2054vs, 2029s, 2005vs, 1976w	1710s	21.4 (3.5), 33.4sh (15)
35a	2043vs, 1978s, 1972s, 1958vs	1730m	20.2 (4.2), 28.3 (6.7)
35e	2041vs, 1974s, 1967s, 1957vs	1730m	20.4 (4.4), 28.9 (7)
35f	2032vs, 1965s, 1959s, 1946vs	1730vs	19.6 (4.8), 29.1 (7)
35c	2042vs, 1980sh, 1975s, 1957vs	1730m	19.5 (3.8), 29.0 (6.1)
35d	2054vs, 1995s, 1988s, 1982vs	1730, 1710m	20.2 (5.0), 28.3 (4.3)
35g	2057vs, 1992s, 1980vs (br)	1740m, 1710s	20.3 (5.0), 28.5 (4.3)
38a	2034vs, 1955vs (br), 1921m		21.9 (4.1), 29.85 (7.41), 33.5sh (6.33)
38b	2028vs, 1950vs, 1945vs		20,9 (4.6), 30.3 (9.5) [5])
38c	2035vs, 1965vs, 1950vs		20.8 (5.5), 29.5 (8.2)
38d	2052vs, 1982vs, 1974vs	1710m	21.7 (4.5), 29.3 (4.6)
36i	2003vs, 1946vs [4])	1765s, 1730s	22.0sh (5.2), 28.5sh (10) [3])
36h	1998vs, 1940vs [5])	1730sh, 1690m	24.0sh (5.2), 28.5sh (9.5), 33.5 (42) [3])
36k	2025vs, 1968vs [5])	1730s, 1690s	20.1sh (2), 22.1 (2.56), 25.6 (3.1) [3])
36l	2021vs, 1960vs [5])	1725s, 1685s	20.9sh (2.2), 22,2 (2.4), 30.3 (8)
36m	1918vs [6])	1720s, 1680s	19.0 (1.9), 27 (18.0)
39e	1997vs, 1940vs [3])	1720	18.5sh (0.69), 24.4 (5.6), 27.8sh (7.9), 35 (20) [6])
39f	1995vs, 1933vs [5])	1710	

[1]) IR in n-hexane, UV in benzene.
[2]) Chloroform solution, −25 °C.
[3]) In n-hexane.
[4]) KBr pellet.
[5]) Dichloromethane solution.
[6]) Toluene solution.

The double-addition products of type XVI are surprisingly stable. Complex *36i, e.g.*, can be crystallized from CH_2Cl_2/MeOH in the presence of hydrochloric acid without substantial decomposition. The most striking property of *36i* is the conversion into the cyclobutadiene complex *33* (R = Ph), which is more conveniently prepared from *28a* and PhC_2Ph at 150 °C. As already mentioned, the reaction proceeds by a stepwise mechanism through complexes XIV–XVI. This reaction offers a new, facile preparation of *33*. The cyclo-dimerisation of alkynes other than diphenylacetylene could not been substantiated, however.

2. Photochemical Cycloadditions of Alkynes and 1.3-Dienes to Diazaferroles

The ability of a transition metal to change symmetry and energy restrictions imposed on cycloaddition reactions is of great preparative and theoretical interest[97, 98]. Contrary to organic cycloadditions, the reactions of coordinated ligands have not been as extensively studied[99–103]. The observation that irradiation of the diazaferrole *35g* in the presence of diphenylacetylene (DPA) or 2.3-dimethylbutadiene yields the cycloadducts *36h* or *42*, prompted a detailed investigation of the mechanism of these reactions; for comparison, simple substitution by PPh_3 (TPP) was also studied[104].

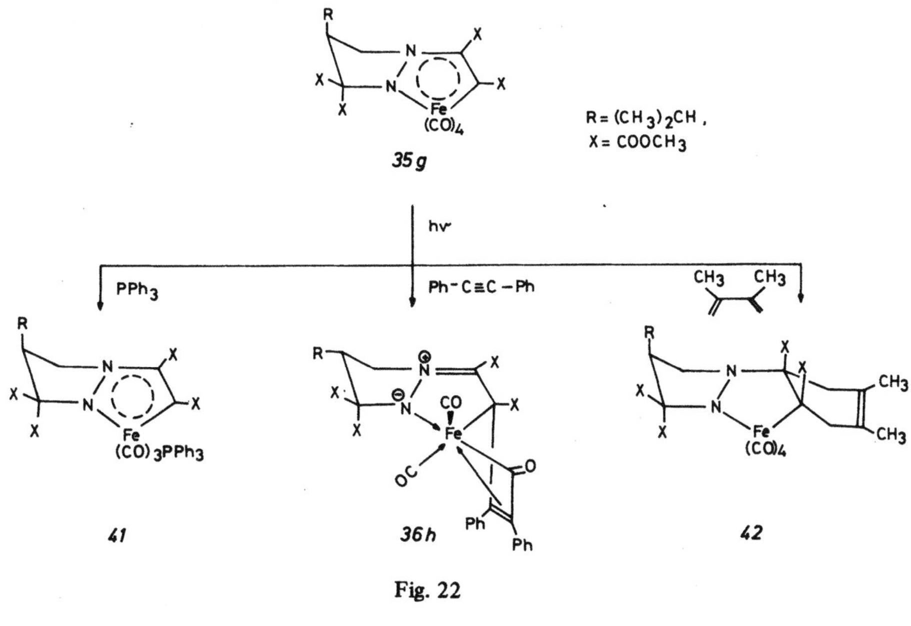

Fig. 22

The reaction with DPA may be viewed as "1.3-dipolar" cycloaddition of the alkyne with the C=Fe=CO moiety of the intermediate formed by loss of one CO group. The DMB-reaction may be considered as Diels-Alder reaction of the cyclic C=C bond.

Flash photolysis of *35g* ((L)$Fe(CO)_4$) yields a rather longlived transient (τ = 70 ms, τ_{max} = 17.8 kK, degassed benzene) which was identified as the coordinatively

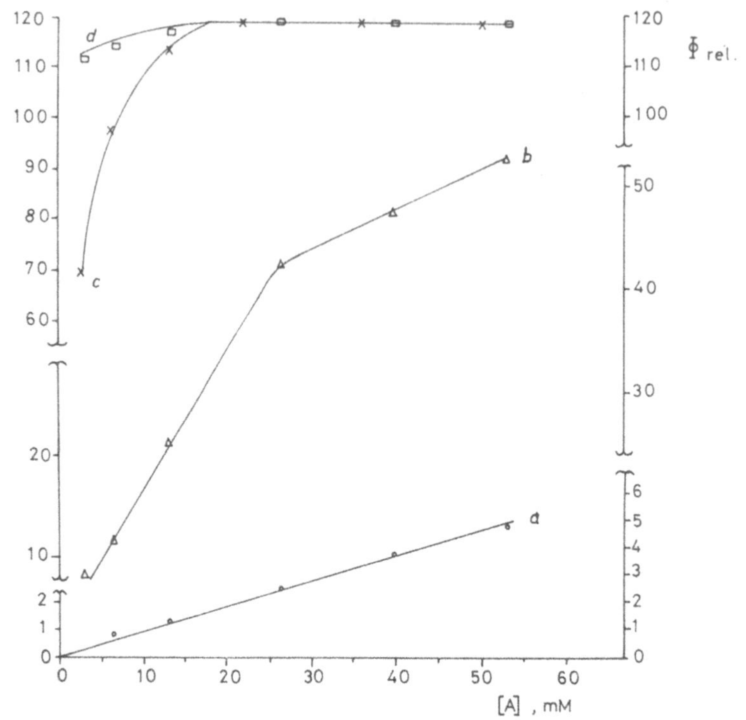

Fig. 23a: Dependence of relative quantum yields on ligand concentration for the reactions of *35g* with DPA (a, b: CO-saturated, degassed solution) and with TPP (c, d: CO-saturated, degassed solution); A refers to TPP or DPA.

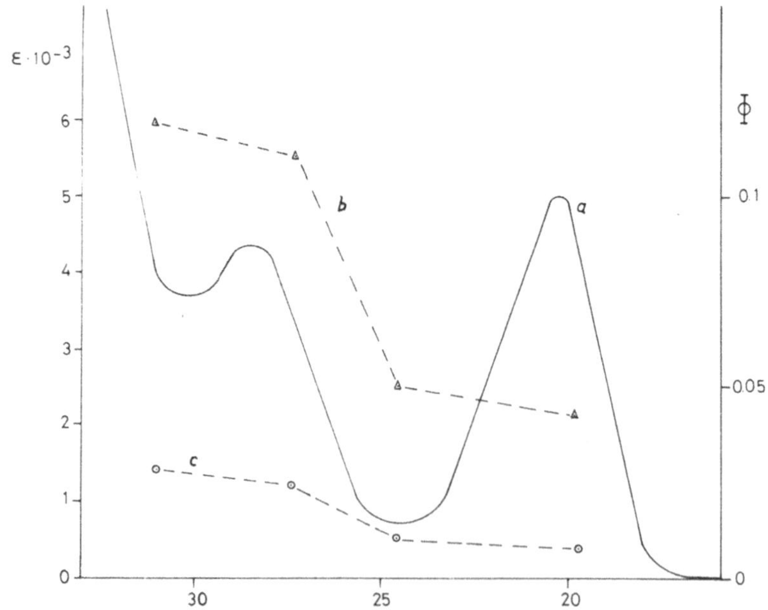

Fig. 23b: Electronic spectrum of complex *35g* (a) and wavelength-dependence of quantum yields of the reaction with TPP (b) and DPA (c)

unsaturated spezies $(L)Fe(CO)_3$ by its reactions with CO or TPP to afford *35g* or *41*, respectively; k_3 (CO) = 2.1×10^5, k_2 (TPP) = 1.3×10^6 $1.mol^{-1}.s^{-1}$. Flash irradiation of *35g* in the presence of DPA or DMB shows again the same transient, from which a longer-lived species is produced showing a new absorption band at 15.5 kK. The rate constants could not be measured in these cases because of overlapping absorption bands. Ligand concentration studies in degassed and CO-saturated solutions reveal that the relative quantum yields of product formation are more strongly dependent on these variables for the DPA and DMB-reaction than for TPP-substitution (Fig. 23 a). The most distinctive feature of the DMB-reaction is the strong dependence on the intensity of the exciting irradiation indicating two or more photochemical reaction steps. Further support for the assumption that these reactions proceed through the common intermediate $(L)Fe(CO)_3$ stems from the qualitatively same dependence of TPP and DPA-reaction on the irradiation-wavelength (Fig. 23 b).

The simplest scheme to account for the experimental results is the following (A refers to TPP or DPA):

$$(L)Fe(CO)_4 \xrightarrow{I_{abs.}} \{(L)Fe(CO)_4\}^* \qquad (1)$$

$$\{(L)Fe(CO)_4\}^* \xrightarrow{k_1} (L)Fe(CO)_3 + CO \qquad (2)$$

$$\{(L)Fe(CO)_4\}^* \xrightarrow{k_{-1}} (L)Fe(CO)_4 \qquad (3)$$

$$(L)Fe(CO)_3 + A \xrightarrow{k_2} (L)Fe(CO)_3A \qquad (4)$$

$$(L)Fe(CO)_3 + CO \xrightarrow{k_3} (L)Fe(CO)_4 \qquad (5)$$

Applying the steady-state approximation for excited state $\{(L)Fe(CO)_4\}^*$ and intermediate $(L)Fe(CO)_3$, the rate of product formation is:

$$\frac{dx}{dt} = \frac{d[(L)Fe(CO)_3A)]}{dt} = \frac{k_1 k_2 I_{abs}[A]}{(k_1 + k_{-1})(k_2[A] + k_3[CO])} \qquad (6)$$

In agreement with the ligand concentration studies, the following simplified cases are considered:

(a) If k_2 [A] $\gg k_3$ [CO] Eq. (6) becomes

$$\frac{dx}{dt} = \frac{k_1}{k_1 + k_{-1}} \cdot I_{abs} \qquad (7)$$

and the quantum yield of product formation is given by

$$\Phi_a = \frac{k_1}{k_1 + k_{-1}} \qquad (8)$$

This case applies to TPP-substitution if [TPP] $\geqslant 3$ [*35g*].

(b) If k_2 [A] $\ll k_3$ [CO] Eq. (6) becomes

$$\frac{dx}{dt} = \frac{k_1 k_2 [A] I_{abs}}{(k_1 + k_{-1}) k_3 [CO]} \tag{9}$$

and

$$\Phi_b = \Phi_a \frac{k_2[A]}{k_3[CO]} \tag{10}$$

The linear increase of Φ_b with increasing [A] [Eq. (10)] is observed in the DPA-reaction in CO-saturated solutions. With the values of Fig. 23a and k_3 as obtained by flash-photolysis, it follows from equ. 10 that k_2(DPA) = $1.2 \times 10^3 l\ mol^{-1} s^{-1}$.

Figure 24 summarizes the mechanisms of these reactions. The excited state $\{(L)Fe(CO)_4\}^*$ more effectively loses CO upon irradiation in the high energy absorption band as indicated by the ratio of k_1/k_{-1} which is 1/9 or 1/20 at 28 or 20 kK, respectively [Eq. (8), Fig. 23b][105]. This is in agreement with predominantly "*d–d*"-character of the high-energy transition and CTLM-character of the lower one, as indicated by SCFF-MO calculations[106]. The rate constant for recombination with CO ($k_3 = 2.1 \times 10^5 l\ mol^{-1} s^{-1}$) compares very well with that observed for $Cr(CO)_5$ + CO $\rightarrow Cr(CO)_6$ ($k = 3 \times 10^6 l\ mol^{-1} s^{-1}$)[107]. The coordinatively unsaturated intermediate $(L)Fe(CO)_3$, common to all three reactions, seems to gain some stability by accepting electrons from the diazaferrole ring as illustrated by the valence bond structures in Fig. 24. This intermediate reacts thermally with TPP or DPA while the DMB-reaction proceeds only photochemically. If low-temperature irradiations are interrupted and the solutions warmed up to room temperature, end-products are only formed in the case of the TPP and DPA-reaction, while no product (*42*) is obtained in the DMB-reaction. Infrared analysis of these cold solutions clearly indicates complex formation of DPA and DMB with $(L)Fe(CO)_3$ before undergoing C–C bond formation. The coordinated alkyne still has one of the degenerate π-orbitals available for intramolecular cycloaddition, while the coordinated diene uses its Ψ_2-orbital for complexation and therefore cannot undergo the cycloaddition reaction by a thermal pathway[89].

From these results it follows obviously that the "Diels-Alder" reaction does not proceed by direct attack of DMB on the cyclic C=C bond but in several distinct steps. This agrees with the formulation of a delocalized ground state for the diazaferroles resulting in a low double bond character. Complex *42* is rather unstable in solution decomposing with cleavage of N–C, Fe–C, and Fe–N bonds into the pyrazoline ligand, 1.2-bis(methoxycarbonyl)-4.5-dimethyl-1.4-cyclohexadiene and iron carbonyl fragments. This instability of *42*, wherein electron delocalization within the five-membered metallocycle is no longer possible, may be taken as chemical evidence for the significant contribution of this electronic effect to the stability of 1.2.3-diazaferroles.

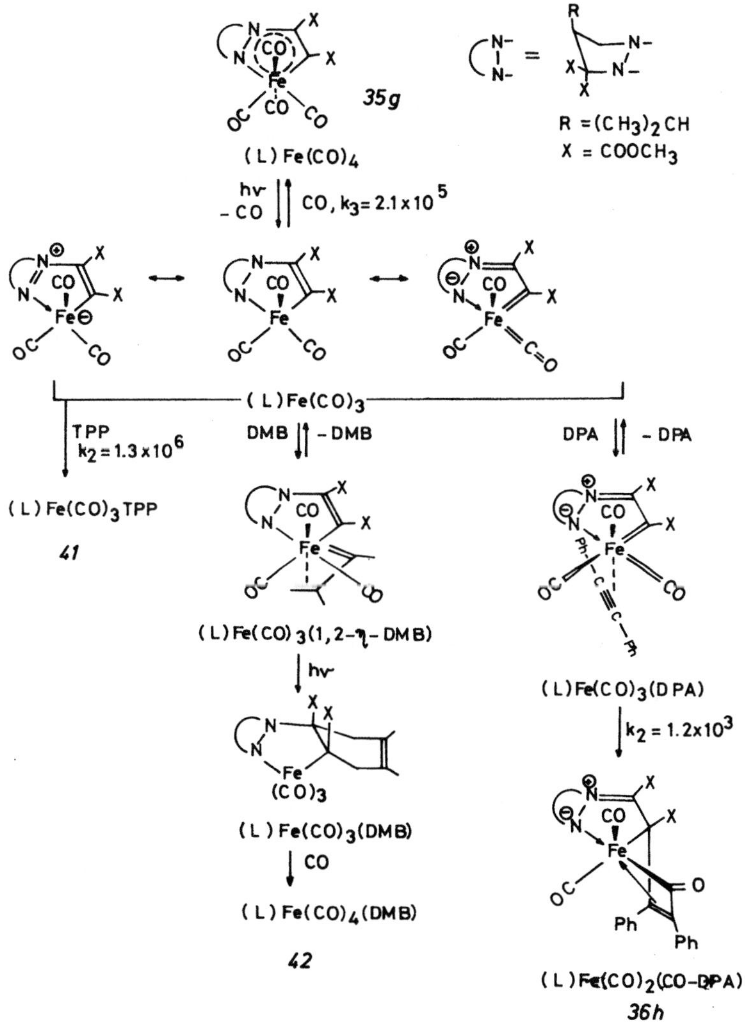
35g
(L) Fe(CO)4
R = (CH3)2CH
X = COOCH3
hν
- CO
CO, k3 = 2.1 x 10^5
(L)Fe(CO)3
TPP
k2 = 1.3 x 10^6
DMB
- DMB
DPA
- DPA
(L) Fe(CO)3TPP
41
(L)Fe(CO)3(1,2-η-DMB)
hν
(L) Fe(CO)3(DMB)
CO
(L) Fe(CO)4(DMB)
42
(L)Fe(CO)3(DPA)
k2 = 1.2x10^3
(L)Fe(CO)2(CO-DPA)
36h

Fig. 24

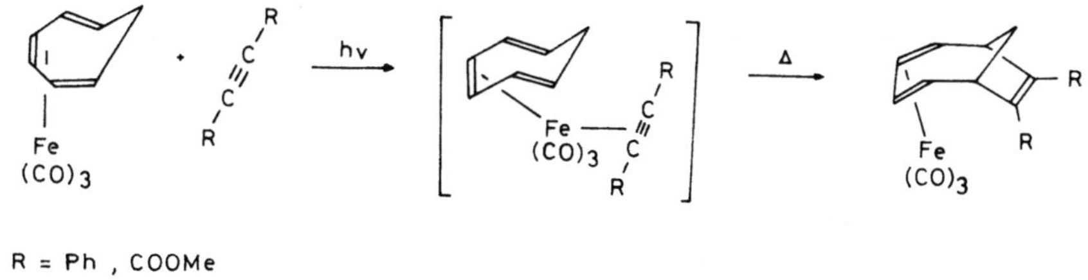
hν
Δ
R = Ph , COOMe

Fig. 25

This stepwise mechanism of the DPA and DMB-reaction agrees very well with the $[\pi_6 + \pi_2]$ cycloaddition of alkynes to tricarbonyl-(cycloheptatriene)-iron and related complexes[99]. Again, molecules without a degenerate pair of highest occupied orbitals react by a different pathway. A step-wise Diels-Alder reaction of hexafluorobut-2-yne, followed by a regiospecific 1.3-hydrogen shift, was performed by the following reaction sequence[101]. The mechanism of these reactions has not been explored in detail.

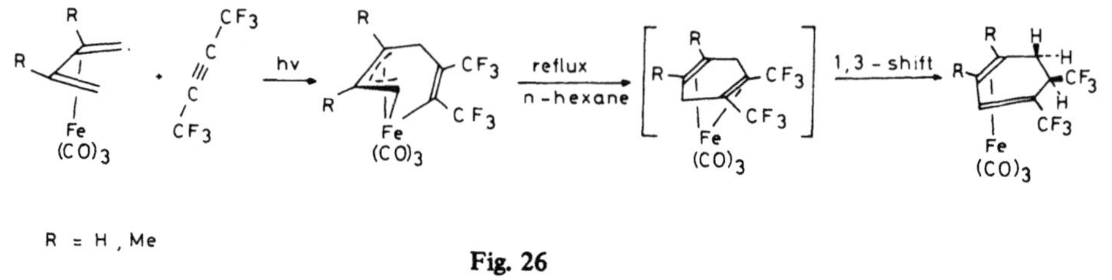

Fig. 26

VI. Complexes Derived from Diazo Compounds

1. Reactions in which Nitrogen is Retained

In contrast to the well known catalytic decomposition of diazo compounds by transition metals[108)], reactions in which nitrogen is retained have been unknown until recently. Lappert and Poland[109)] reported an interesting 1.3-insertion of trimethyl-silyldiazomethane into M–H bonds of group VI b carbonyls affording the

$$\mathrm{Cp\,M(CO)_3H + N{=}N{=}C(H)R \xrightarrow{-CO} Cp\,M(CO)_2(N_2CH_2R)} \qquad M = Mo, W;\ R = Me_3Si, H \tag{1}$$

hitherto unknown alkyldiazenato complexes [Eq. (1)]. The same type of reaction occurs also with diazomethane, as was shown by Herrmann[110)]. There is some evidence that the reaction is initiated by displacement of CO through the diazoalkane[110)]. The structure of these products is based mainly on the ν(N=N) vibrations of the diazenato ligand[109, 110)] which appear between 1590 and 1640 cm^{-1}. If the corresponding metal carbonyl anions are used in these reactions, the diazoalkane attacks an M = CO bond in the way similar to a "1.3-dipolar" cycloaddition[95, 96)] to give

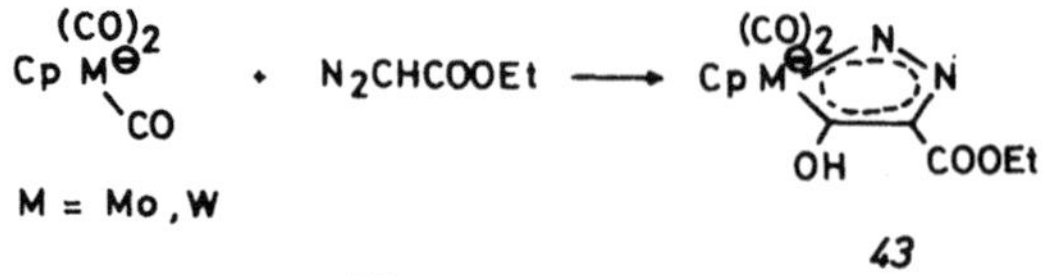

43

Fig. 27

the five-membered metallocycles[93]. Both reactions nicely reflect the "1.3-dipolar" properties of the HOMO of diazo compounds (Fig. 2). A further example of this behaviour is the displacement of dinitrogen by $(CF_3)_2CN_2$ to yield the iridacycles of proposed structure[111] *44*.

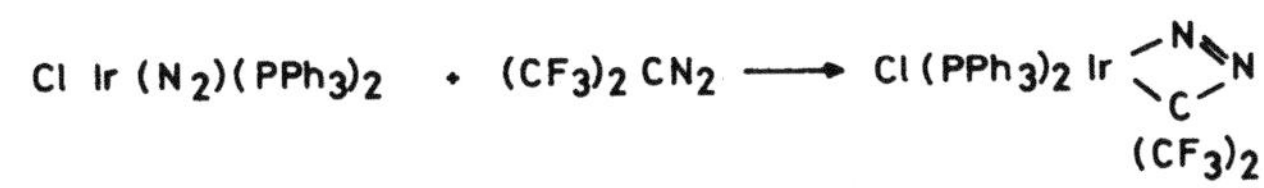

44

Fig. 28

The complexes *43*, *44* are examples of the general types XI and X (Fig. 4).

Otsuka, Nakamura *et al.*[112] prepared the first complexes containing "side-on" coordinated diazo compounds by ligand displacement reactions [Eqs. (2–4)]. The

$$\text{"M(t-BuNC)}_2\text{"} + R_2CN_2 \rightarrow M(t\text{-}BuNC)_2(R_2CN_2) \quad (2)$$

$$M(C_2H_4)(PPh_3)_2 + R_2CN_2 \rightarrow M(PPh_3)_2(R_2CN_2) \quad (3)$$

$$Ni(1.5\text{-}COD)_2 + R_2CN_2 \rightarrow Ni(1.5\text{-}COD)(R_2CN_2) \quad (4)$$

M = Pd, Ni. R = Ph, CN; R_2 = 9-fluorenylidene

proposed structure is supported by reactions of the 9-diazofluorene complex with oxygen and diethylmaleate affording free diazofluorene and the corresponding oxygen and maleate complexes. Thermal decomposition of $Ni(t\text{-}BuNC)_2$(9-diazofluorene) at 100 °C yields fluorenone azine, fluorenone, 9-cyanofluorene, and N-(t-Bu)-9-fluorenylacetamide[112]. The characteristic ν(NNC) vibration of the free diazo compound is lowered by an average of 500 cm^{-1} upon coordination. This indicates strong "back-donation" into the antibonding $3b_2$ orbital which is localized predominantly on the N=N bond (Fig. 2). In agreement with the strong π-acceptor properties of the diazo ligand, the ν(NC) frequencies of the isocyanides are shifted to higher wavenumbers in the corresponding complexes and compare very well with the values of azobenzene and maleic anhydride complexes (Table 12). From these results and the fact that the frontier orbitals of a diazo compound are very similar to ketenes and ketenimines[26], a "side-on" coordination of the diazo ligand was formulated (XVII) by analogy with the known structure of $Ni(t\text{-}BuNC)_2$ $(t\text{-}BuN{=}C{=}C(CN)_2)$[113] (*45*).

Structures of type XVIII or XIX as proposed for the diazodicyanomethane complex (Table 12) may also apply to the complexes of diphenyldiazomethane and 9-diazo-fluorene. The formation of the ketenimine complex (*45*) from the reaction of $(CN)_2CN_2$ and $Ni(t\text{-}BuNC)_4$ probably occurs via attack of the complexed dicyanomethylene carbene on the isocyanide ligand[113]. The observation that these complexes[112, 113] catalyze the formation of ketenimines from isocyanides and diazo compounds, a reaction which does not proceed under same conditions without the transition metal, may be of preparative value[113].

Table 12. Characteristic IR absorptions of diazo- and related complexes

Complex	ν(N=C)[1]	ν(C=N=N)[1]	Ref.
Ni (t-BuNC)$_2$ (R$_2$CN$_2$)	2180	1509	[112]
Ni (PPh$_3$)$_2$ (R$_2$CN$_2$)	–	1455	[112]
Pd (t-BuNC)$_2$ (R$_2$CN$_2$)	2190 2170	1547	[112]
Pd (PPh$_3$)$_2$ (R$_2$CN$_2$)	–	1514	[112]
Ni (t-BuNC)$_2$ (N$_2$C(CN)$_2$)	2200 2180	1550	[113]
Ni (t-BuNC)$_2$ (Ph$_2$CN$_2$)	2169 2139	1488	[112]
Ni (PPh$_3$)$_2$ (Ph$_2$CN$_2$)	–	1453	[112]
Ni (COD) (R$_2$CN$_2$)	–	1521	[112]
Ni (t-BuNC)$_2$ (Ph-N=N-Ph)	2168 2140	–	[64]
Ni (t-BuNC)$_2$ (maleic anhydride)	2159 2118	– –	[64]

[1]) Measured in Nujol mull; the frequencies of free t-BuNC and diazo compounds are at 2130 and 2000–2150 cm^{-1}. R$_2$ = 9-fluorenylidene.

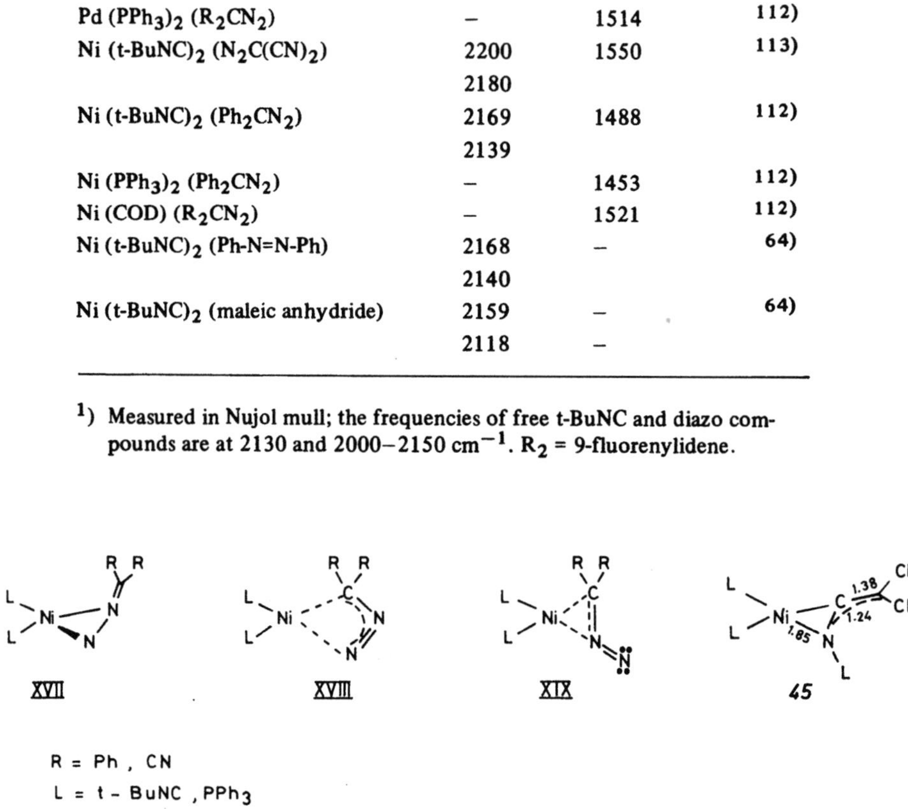

Fig. 29

Complexes which probably contain a diazo ligand coordinated in an "end-on" fashion were obtained via displacement reactions of Cp Mn(CO)$_2$(THF)[114]. Note that the organic by-product is not the expected tetra-(ethoxycarbonyl)ethylene but its hydrogenated derivative *47*.

Phenylmethyldiazomethane does not react analogously but yields the ketimine complex CpMn(CO)$_2$(PhMeC=NH) via cleavage of the N=N bond (see VI. 2)[87]. The nature of the Mn–N bond in *46* should be similar to that of the dinitrogen complex CpMnN$_2$(CO)$_2$ since the highest occupied σ-orbital of a diazo compound (7a_1) and of dinitrogen closely resemble each other[26].

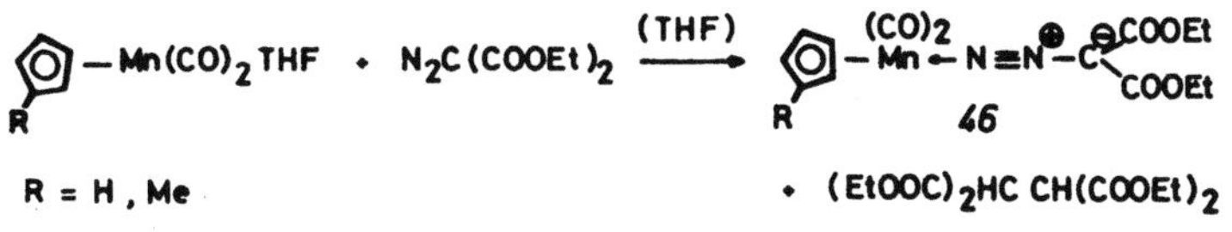

Fig. 30

Fig. 31

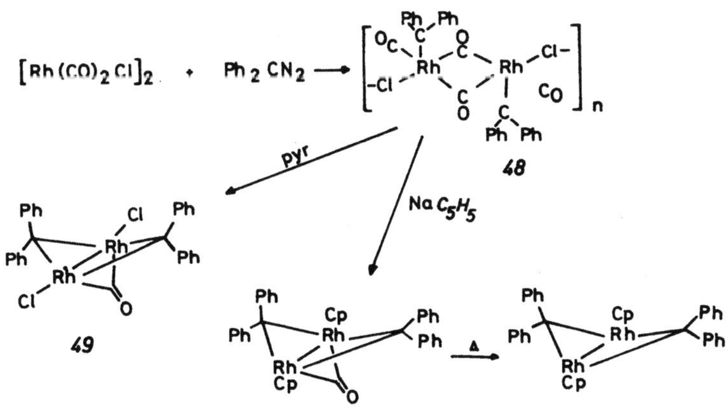

Fig. 32

2. Formation of Carbene Complexes

It is significant that the preparation of carbene complexes from neutral organic precursers has proved largely unsuccessful[115)]. However, recently the synthesis of several complexes was accomplished by using diazo compounds as carbene generating systems. Reaction of $[Rh(CO)_2Cl]_2$ with diphenyldiazomethane produces[116)] the polymeric complex *48*, which is converted into the compounds *49* or *50* by treatment with pyridine or NaC_5H_5 respectively. The structures of *49–51* were established by X-ray analysis[117)]. Hermann[88)] prepared a number of mononuclear carbene complexes by reaction of $CpMn(CO)_2(THF)$ with diphenyldiazomethane derivatives in the presence of copper powder (Fig. 33, 34). The carbene ligand of complexes *52* can be cleaved off by ammonium cerium (IV) nitrate or thermolysis to give benzophenone or tetraphenylethylene derivatives, respectively. It is interesting that the reaction with

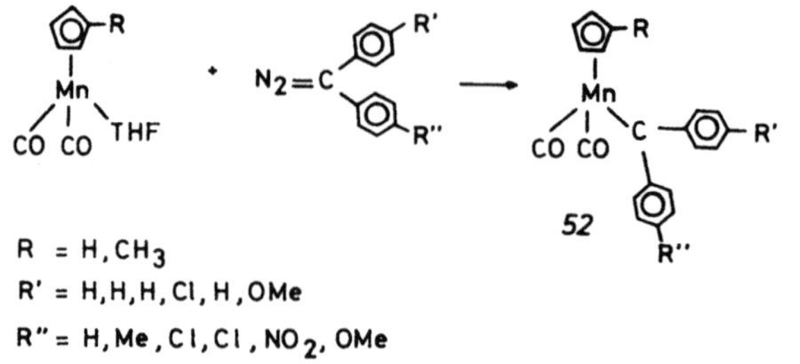

Fig. 33

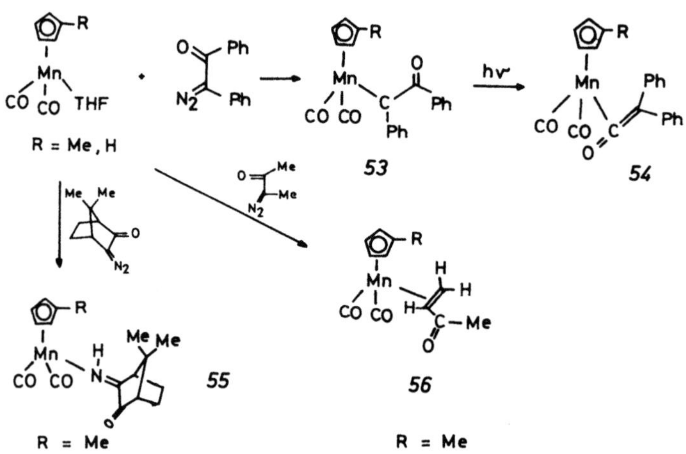

Fig. 34

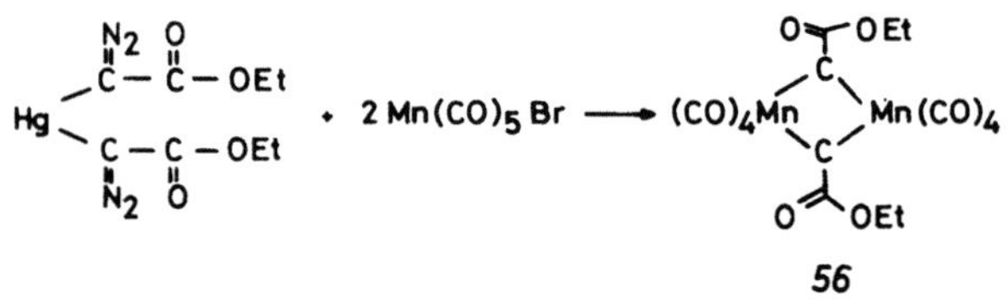

Fig. 35

benzoylphenyldiazomethane does not lead to complexes of diphenylketene but yields the corresponding ketocarbene complex *53*. Thus, the Wolff-rearrangement of the diazoketone is stopped after its rate-determining step of nitrogen evolution. The rearrangement of the complexed ketocarbene can be effected by uv-irradiation affording the diphenylketene complexes[88] *54*. 3-diazo-2-butanone reacts by an intramolecular hydride transfer to give *56* while in the case of 3-diazocamphor cleavage of the N=N bond yields the 3-iminocamphor complex *55*. In addition to the synthesis of carbene complexes it was also possible to prepare the bridged "ethoxycarbonylcarbyne" complex *56* by reaction of $MnBr(CO)_5$ with $Hg[C(N_2)COOEt]_2$ in refluxing ether[118, 119] (Fig. 35).

VII. Conclusion

It is evident from the results presented above that the cis-(–N=N–) bond has powerful coordinating properties towards low valent transition metals. Due to the unique electronic structure of the diazeno group, many different modes of interaction have been observed and it seems promising that this ability of the N=N function may be utilized in catalytic processes. Since diazene itself is a proposed intermediate in biological nitrogen-fixation, the observed modes of coordination are of some relevance to this problem. The results obtained from the cycloaddition reactions are of general importance for the investigation of the factors determining photochemical or thermal reactions of coordinated ligands. The ability of iron carbonyls to activate the diazeno group towards attack by unsaturated molecules like alkynes, offers a convenient synthesis of novel organic heterocycles. Together with the recently reported conversion of coordinated dinitrogen into diazeno and diazenato complexes[4)], these reactions may be key steps in the incorporation of N_2 into organic compounds.

Acknowledgement: We wish to thank Dr. D. Henneberg (mass spectra), Dr. K. Hildenbrandt (^{13}C-NMR spectra), Dr. C. Krüger (X-ray analysis), Dr. F. Mark and Dr. F. Olbrich (MO calculations), H. Matthäus, A. Riemer, B. Ulbrich (technical assistance), and the Humboldt-Stiftung for a postdoctoral fellowship to one of us (A. A.). Generous support by Prof. O. E. Polansky and Prof. G. O. Schenck is gratefully acknowledged.

VIII. References and Notes

1) For a comprehensive summary see: Herberhold, M.: Metal π-complexes. Amsterdam–New York–London: Elsevier Publ. Company 1972.

2) a) Bruce, M. I., and Goodall, B. L. in: The chemistry of hydrazo, azo and azoxy groups (ed. S. Patai) Vol. I, Chap. 9, p. 259. New York: John Wiley 1975;
b) Kilner, M.: Adv. Organometal. Chem. *10*, 115 (1972);
c) Carty, A. J.: Organometal. Chem. Rev. A. 7, 191 (1972);
d) Sutton, D.: Chem. Soc. Rev. *4*, 443 (1975).

3) See *e.g.*: Postgate, J. R., (ed.): The chemistry and biochemistry of nitrogen fixation. London and New York: Plenum Press 1971;
Schrauzer, G. N.: Angew. Chem. *87*, 579 (1975);
Angew. Chem. internat. Edit. *14*, 514 (1975).

4) Chatt, J.: J. Organometal. Chem. *100*, 17 (1975);
Day, V. W., George, T. A., and Iske, S. D. A.: J. Amer. Chem. Soc. *97*, 4127 (1975).

5) The following abbreviations are used: THF = tetrahydrofurane, Cp = η^5-C_5H_5, COD = 1,5-cyclooctadiene, pyr = pyridine, TPP = triphenylphosphine, DPA = diphenylacetylene, DMB = 2.3-dimethylbutadiene, nor-C_7H_8 = norbornadiene.

6) Jaffé, H. H., Si-Jung Ye, and Gardner, R. W.: J. Mol. Spectroscopy *2*, 120 (1958).

7) Hochstrasser, R., and Lower, S.: J. Chem. Phys. *36*, 3505 (1962).

8) Robin, M. B., and Simpson, W. T.: J. Chem. Phys. *36*, 580 (1962).

9) a) Robin, M. B., Hart, R. R., and Kuebler, N. A.: J. Amer. Chem. Soc. *89*, 1564, (1967);
b) Vasudevan, K., Peyerimhoff, S. D., Buenker, R. J., Kammer, W. E., and Hsiang-lin Hsu: Chem. Phys. 7, 187 (1975);
c) Frost, D. C., Lee, S. T., McDowell, C. A., and Westwood, N. P. C.: Chem. Phys. Letters *30*, 26 (1975).

10) Series of papers, *e.g.*: Brogli, F., Eberbach, W., Haselbach, E., Heilbronner, E., Hornung, V., and Lemal, D. M.: Helv. Chim. Acta *56*, 1933 (1973) and references therein.

11) We feel that this is more appropriate than using "n_-" for the HOMO of both cis and trans-diazenes[12] since "n_-" is always associated with an antisymmetric combination.

12) Houk, K. N., Chang, Y. M., and Engel, P. S.: J. Amer. Chem. Soc. *97*, 1824 (1975). This paper contains an exhaustive summary of hitherto published results of MO-calculations and photoelectron spectra of diazenes.

13) Gimarc, B. M.: J. Amer. Chem. Soc. *92*, 266 (1969).

14) Hoffmann, R.: Acc. Chem. Res. *4*, 1 (1971).

15) Gleiter, R., Heilbronner, E., and Hornung, V.: Angew. Chem. *82*, 878 (1970). Ang. Chem. internat. Edit. *9* 902 (1970).

16) a) Jaffé, H. H., and Orchin, M.: Theory and applications of ultraviolet spectroscopy. New York: John Wiley and Sons 1964.
b) Kobayashi, T., Yokota, K., and Nagakura, S.: J. Electron. Spectroscop. *6*, 167 (1975).

17) Potzinger, P., and Sonntag, W.: private communication 1974.

18) Honner, J. R., Nott, P. R., and Selinger, B. K.: Aust. J. Chem. *27*, 1613 (1974).

19) Albini, A., and Kisch, H.: J. Organometal. Chem. *94*, 75 (1975).

20) Robin, M. B., and Simpson, W. T.: J. Chem. Phys. *36*, 580 (1962).

21) Robin, M. B.: Higher excited states of polyatomic molecules, Vol. II, p. 69. New York: Academic Press 1975.

22) Hoffmann, R.: Tetrahedron *22*, 539 (1966).

23) Haselbach, E., and Heilbronner, E.: Helv. Chim. Acta *53*, 684 (1970).

24) Olbrich, F.: private communication 1975.

25) Haselbach, E., Heilbronner, E., Mannschreck, A., and Seitz, W.: Angew. Chem. *82*, 879 (1970); Ang. Chem. internat. Edit. *9*, 902 (1970).

26) Jorgensen, W. L., and Salem, L.: Orbitale Organischer Moleküle, S. 120. Weinheim: Verlag Chemie 1974.

27) Schuster, P., and Polansky, O. E.: Monatsh. Chem. *96*, 396 (1964).

28) Heilbronner, E., and Martin, H.-D.: Chem. Ber. *106*, 3376 (1973).
29) Kisch, H.: J. Organometal. Chem. *38*, C19 (1972).
30) Albini, A., and Kisch, H.: Angew. Chem. *87*, 206 (1975); Internat. Edit. *14*, 182 (1975).
31) Knox, G. R.: private communication 1974.
32) Bagga, M. M., Flannigan, W. T., Knox, G. R., and Pauson, P. L.: J. Chem. Soc. (C) *1969*, 1534.
33) Albini, A., and Kisch, H.: unpublished results.
34) Nakamura, A., Aotake, M., and Otsuka, S.: J. Amer. Chem. Soc. *96*, 3456 (1974).
35) Rybin, L. V., Arutyunyan, A. V., Petrovskii, P. V., and Rybinskaya, M. I.: Bull. Acad. Sci. USSR, Div. Chem. Sci. *21*, 184 (1972).
36) Kisch, H.: unpublished results.
37) Nesmeyanov, A. N., Babin, V. N., Kochetkova, N. S., and Nekrasov, Yu. S.: Dokl. Chem. *200*, 792 (1971).
38) Cotton, F. A., and Troup, J. M.: J. Amer. Chem. Soc. *96*, 3438 (1974).
39) Krüger, C.: Chem. Ber. *106*, 3230 (1973).
40) a) Schubert, E. H., and Sheline, R. K.: Inorg. Chem. *5*, 1071 (1966).
b) For a detailed discussion of the assignment of axial and equatorial positions in $(L)Fe(CO)_4$ complexes see: Darensbourg, D. J., Nelson, III, H. H., and Hyde, C. L.: Inorg. Chem. *13*, 2135 (1974).
41) Alper, H.: J. Organometal. Chem. *50*, 209 (1973).
42) Note that the lengthening of the N=N bond is not necessary a criterion for the presence of π-bonding in these complexes: σ-donation through the n_1 orbital, which is antibonding with respect to the N=N bond, is expected to slightly strengthen this linkage, whereas back-donation should cause the opposite and therefore no net effect may be observed.
43) Weigel, H., and Mark, F.: private communication 1974.
44) Herberhold, M., and Golla, W.: Chem. Ber. *107*, 3199 (1974).
45) Huttner, G., Schmid, H. G., Willenberg, H., and Stark, Th.: J. Organometal. Chem. *94*, C3 (1975).
46) Beck, W., and Danzer, W.: Z. Naturforschg. B, in press (1975).
47) Ackermann, M. N., and Lee-Juan Kou: J. Organometal. Chem. *86*, C7 (1975).
48) Herberhold, M., Golla, W., and Leonhard, K.: Chem. Ber. *107*, 3209 (1974).
49) Herberhold, M., Leonhard, K., Golla, W., and Alt, H.: Abstract of Papers, VI. Int. Conf. Organometal. Chem. Amherst, Mass. 1973, Abstract 117.
50) Herberhold, M., and Golla, W.: J. Organometal. Chem. *26*, C27 (1971).
51) Kooti, M. and Nixon, J. F.: J. Organometal. Chem. *76*, C29 (1974).
52) Cotton, F. A., Edwards, W. T., Rauch, F. C., Graham, M. A., Perutz, R. N., and Turner, J.J.: J. Coord. Chem. *2*, 247 (1973).
53) Herberhold, M., Leonhard, K., and Kreiter, C. G.: Chem. Ber. *107*, 3222 (1974).
54) Snyder, J. P., Bandurco, V. T., Darack, F., and Olsen, H.: J. Amer. Chem. Soc. *96*, 5158 (1974).
55) Van Baar, J. F., Vrieze, K., and Stufkens, D. J.: J. Organometal. Chem. *81*, 247 (1974).
56) Sellmann, D.: J. Organometal. Chem. *44*, C46 (1972).
57) Sellmann, D., Brandl, A., and Endell, R.: Angew. Chem. *85*, 1121 (1973); Angew. Chem. internat. Edit. *12*, 1019 (1973); J. Organometal. Chem. *49*, C22 (1973).
58) Sellmann, D., Brandl, A., and Endell, R.: Angew. Chem. *85*, 1122 (1973); Ang. Chem. internat. Edit. *12*, 1019 (1973).
59) Huttner, G., Gartzke, W., and Allinger, K.: J. Organometal. Chem. *91*, 47 (1975).
60) Sellmann, D., Brandl, A., and Endell, R.: J. Organometal. Chem. *90*, 309 (1975).
61) For reviews on the relation between colour and constitution of diazenes, see: Bock, H.: Angew. Chem. *77*, 469 (1965); Angew. Chem. internat. Edit. *4*, 457 (1965); Rau, H.: *ibid. 85*, 24 (1973); intern. Edit. *12*, 224 (1973).
62) Herberhold, M., and Leonhard, K.: J. Organometal. Chem. *78*, 253 (1974).
63) Otsuka, S., Yoshida, T., and Tatsuno, Y.: Chem. Commun. *1971*, 67.
64) Otsuka, S., Yoshida, T., and Tatsuno, Y.: J. Amer. Chem. Soc. *93*, 6462 (1971).
65) Klein, H.-F., and Nixon, J. F.: Chem. Commun. *1971*, 42.

66) Klein, H.-F.: Angew. Chem. *83*, 363 (1971), Angew. Chem. internat. Edit. *10*, 343 (1971).
67) Fachinetti, G., Fochi, G., and Floriani, C.: J. Organometal. Chem. *57*, C51 (1973).
68) McDonald, J. W., Newton, W. E., Creedy, Ċ. T., and Corbin, J. L.: J. Organometal. Chem. *92*, C25 (1975).
69) Nakamura, A., Aotake, M., and Otsuka, S.: J. Amer. Chem. Soc. *96*, 3456 (1974).
70) Ittel, St. D., and Ibers, J. A.: Inorg. Chem. *14*, 1183 (1975).
71) Ittel, St. D., and Ibers, J. A.: J. Organometal. Chem. *57*, 389 (1973).
72) Dickson, R. S., and Ibers, J. A.: J. Amer. Chem. Soc. *94*, 2988 (1972).
73) Ittel, St. D., and Ibers, J. A.: J. Organometal. Chem. *74*, 121 (1974).
74) Brown, C. J.: Acta Crystallogr. *21*, 146 (1966).
75) It is noted that this may not be a "pure" (N=N) vibration since extensive coupling with vibrational modes of the phenyl ring occurs in several aryldiazenato complexes: Haymore, B. L., Ibers, J. A., and Meek, D. W.: Inorg. Chem. *14*, 541 (1975).
76) Ittel, St. D., and Ibers, J. A.: J. Organometal. Chem. *74*, 121 (1974).
77) Bennett, R. P.: Inorg. Chem. *9*, 2184 (1970).
78) Kisch, H.: J. Organometal. Chem. *30*, C25 (1971).
79) Nesmeyanov, A. N., Rybin, L. V., Rybinskaya, M. I., Arutyunyan, A. V., Gubenko, N. T., and Petrovski, P. V.: Bull. Acad. Sci. USSR, Div. Chem. Sci. *20*, 1478 (1971).
80) Albini, A., and Kisch, H.: J. Organometal. Chem. 1975, in press.
81) Mastropasqua, P., Riemer, A., and Kisch, H.: unpublished results.
82) Note that the diazomethane obtained by irradiation of diazirine is not formed via isomerisation but by recombination of $CH_2 + N_2$; Pimentel, G. C. and Moore, C. B.: J. Chem. Phys. *41*, 3504 (1964). However, in the case of 3-Aryl-3H-Diazirines direct isomerisation is possible; Smith, R. A. G. and Knowles, J. R., Perkin II, J. C. S.: *1975*, 686.
83) There exist some statements that the coordination behaviour of diazirines should be very distinct from that of other diazenes[84)].
84) Frey, H. M.: Adv. Photochem. *4*, 255 (1966); Mackenzie, K., in: The chemistry of hydrazo, azo and azoxy groups (ed. S. Patai), Vol. I, p. 335. New York: John Wiley 1975.
85) Krüger, C.: private communication, 1974.
86) Ashley-Smith, J., Green, M., Mayne, N., and Stone, F. G. A.: Chem. Comm. *1969*, 409.
87) Herrmann, W. A.: Chem. Ber. *107*, 2899 (1974).
88) Herrmann, W. A.: Chem. Ber. *108*, 486 (1975).
89) Assuming that the possibility of a chemical reaction is determined by favourable HOMO-LUMO interactions of the reacting species (FMO-Theory)[90] and neglecting steric factors.
90) Fukui, K.: Topics in Current Chemistry, *15*, 1 (1970).
91) Krüger, C., and Kisch, H.: J. C. S. Chem. Commun. *1975*, 65.
92) Hildenbrandt, K.: private communication, 1974.
93) Green, M. L. H., and Sanders, I. R.: J. Chem. Soc. (A) *1971*, 1947.
94) For a recent review see: Yur' eva, L. P.: Russ. Chem. Rev. *43*, 48 (1974).
95) A 1.3-dipol is defined as an organic compound containing four π-electrons distributed among three atoms[96)].
96) a) Huisgen, R.: Angew. Chem. *75*, 604 (1963), Int. Ed. *2*, 565 (1963).
b) Gilchrist, T. C., and Storr, R. C.: Organic reactions and orbital symmetry, p. 111. Cambridge: University Press 1971.
97) For a recent review on catalytic cyclooligomerisation reactions see *e.g.*: Heimbach, P.: Angew. Chem. *85*, 1035 (1973), Angew. Chem. internat. edit. *12*, 975 (1973).
98) For theoretical treatments see: a) Pearson, R. G.: Topics in Current Chemistry *1974*, 75 and references therein;
b) Mango, F. D.: ibid. p. 39;
c) Heimbach, P., and Traunmüller, R.: Metall-Olefin-Komplexe, Weinheim: Verlag Chemie 1970;
d) Mango, F. D.: Coord. Chem. Rev. *15*, 109 (1975).
99) Davis, R. E., Dodds, T. A., Hseu, T.-H., Wagnon, J. C., Devon, T., Tancrede, J., McKennis, J. S., and Pettit, R.: J. Amer. Chem. Soc. *96*, 7562 (1974).
100) Patel, H. A., Carty, A. J., Mathew, M., and Palenik, G. J.: J. C. S. Chem. Comm. *1972*, 810.

101) Davis, R., Green, M., and Hughes, R. P.: J. C. S. Chem. Comm. *1975*, 405.
102) Bottrill, M., Goddard, R., Green, M., Hughes, R. P., Lloyd, M. K., Lewis, B., and Woodward, P.: J. C. S. Chem. Comm. *1975*, 253.
103) Davidson, J. L., Green, M., Stone, F. G. A., and Welch, A. J.: J. C. S. Chem. Comm. *1975*, 286.
104) Albini, A., and Kisch, H.: J. Amer. Chem. Soc. 1975, in press.
105) For a recent review on the photochemistry of metal carbonyls see: Wrighton, M.: Chem. Rev. *1974*, 401.
106) Weigel, H., and Mark, F.: private communication 1975.
107) Kelly, J. M., Hermann, H., and Koerner von Gustorf, E.: J. C. S. Chem. Comm. *1973*, 105.
108) Kirmse, W.: Carbene chemistry, 2nd edit. New York, London: Academic Press, Moser, W. R.: J. Amer. Chem. Soc. *91*, 1135, 1141 (1969); Wulfman, D. S., Peace, B. W., and Steffen, E. K.: J. C. S. Chem. Comm. *1971*, 1360; Peace, B. W., and Wulfman, D. S.: J. C. S. Chem. Comm. *1971*, 1179;
Paulissen, R., Hubert, A. J., and Theyssie, Ph.: Tetrahedron Lett. *1972*, 1465.
109) Lappert, M. F., and Poland, J. S.: Chem. Comm. *1969*, 1061.
110) Herrmann, W. A.: Angew. Chem. *87*, 358 (1975), Angew. Chem. internat. Edit. *14*, 355 (1975).
111) Clemens, J., Green, M., and Stone, F. G. A.: J. C. S. Dalton *1973*, 1620.
112) Otsuka, S., Nakamura, A., Koyama, T., and Tatsuno, Y.: Liebigs Ann. Chem. *1975*, 626.
113) Yarrow, D. J., Ibers, J. A., Tatsuno, Y., and Otsuka, S.: J. Amer. Chem. Soc. *95*, 8590 (1973).
114) Herrmann, W. A.: J. Organometal. Chem. *84*, C25 (1975).
115) Cardin, D. J., Cetinkaya, B., and Lappert, M. F.: Chem. Rev. *72*, 545 (1972).
116) Hong, P., Nishi, N., Sonogashira, K., and Hagihara, N.: J. C. S. Chem. Comm. *1972*, 993.
117) Yamamoto, T., Garber, A. R., Wilkinson, J. R., Boss, Ch. B., Streib, W. E., and Todd, L. J.: J. C. S. Chem. Comm. *1974*, 354.
118) Herrmann, W. A.: Angew. Chem. *86*, 895, (1974), Angew. Chem. internat. Edit. *13*, 812 (1974).
119) For a definition of a carbyne complex see: Fischer, E. O., and Schubert, U.: J. Organometal. Chem. *100*, 59 (1975).

Received November 27, 1975

Author Index Volumes 26–65

The volume numbers are printed in italics

Molekülverbindungen und Koordinationsverbindungen in Einzeldarstellungen

Herausgeber: G. Briegleb, F. Cramer, H. Hartmann

G. Briegleb:
Elektronen-Donator-Acceptor-Komplexe
Unter Mitwirkung von W. Liptay.
93 Abbildungen. X, 279 Seiten. 1961

Ch. K. Jørgensen:
Oxidation Numbers and Oxidation States
VII, 291 pages. 1969

H.-H. Perkampus:
Wechselwirkung von Pi-Elektronensystemen mit Metallhalogeniden
64 Abbildungen. 37 Tabellen. XI, 215 Seiten. 1973

H. L. Schläfer:
Komplexbildung in Lösung
Methoden zur Bestimmung der Zusammensetzung und der Stabilitätskonstanten gelöster Komplexverbindungen.
76 Abbildungen. X, 348 Seiten. 1961

Springer-Verlag
Berlin
Heidelberg
New York

GPSR Compliance

The European Union's (EU) General Product Safety Regulation (GPSR) is a set of rules that requires consumer products to be safe and our obligations to ensure this.

If you have any concerns about our products, you can contact us on ProductSafety@springernature.com

In case Publisher is established outside the EU, the EU authorized representative is:

Springer Nature Customer Service Center GmbH
Europaplatz 3
69115 Heidelberg, Germany

Batch number: 10620881

Printed by Printforce, the Netherlands